Cool Fusion

A Quantum Approach to Peak Minerals, Nuclear Waste, and Future Metals Shock

Edward Esko
and Alex Jack

Foreword by Dr. Mahadeva Srinivasan

Revised Edition

Cool Fusion:
A Quantum Approach to Peak Minerals, Nuclear Waste, and Future Metals Shock
Copyright © 2012 by Edward Esko and Alex Jack
ISBN-10: 1477563725
ISBN-13: 978-1477563724

All rights reserved. Printed in the United States of America. No part of this book may be used or reproduced in any manner whatsoever, including digital, without written permission except in the case of brief quotations embodied in critical articles or reviews.

For further information on mail-order sales, wholesale or retail discounts, distribution, translations, and foreign rights, please contact the publisher:

Amber Waves
P.O. Box 487
Becket MA 01223
Tel (413) 623-0012
Email: shenwa@bcn.net

First Edition: June 2011
Second Edition: June 2012

Cover: Sunrise over Owls Head, ME, Copyright © Edward Esko.

Quantum Rabbit Founders Alex Jack (left) and Edward Esko in Massachusetts, 2005. Photo: Julia Esko

The experiments described here are done in simple laboratories with a low budget, yet are very significant because they show with very little doubt that elements can be transmuted at temperatures far below what conventional science would consider possible. Other experiments over decades have shown similar results by different methods but have been ignored by the general scientific community. These results should be harder to ignore because some of the transmuted elements showed up in quantities that far exceed what would be possible taking the worst-case errors in the process. Congratulations to Edward and the team. — Bill Zebuhr, editor *Infinite Energy* magazine

As remote as the possibility of LENR of heavier elements may seem, if it was confirmed it would have earth-shaking consequences, both theoretically and practically. Being in conflict with the current theories of physics, it would possibly pave the way for a new revolution that would likely bring about a completely new understanding of physics. It would surely open the door for a huge number of applications that could introduce radically new industries, like producing iron from coal or copper from iron and lithium. Further exploration could well uncover new inexhaustible sources of energy from nuclear reactions that could be exploited in safe and peaceful ways. — Matthias Grabiak, physicist

What we're doing is hotter than cold fusion, but a lot colder than hot fusion. It's within the range of temperatures found on earth. We use the term 'cool fusion' to describe it. — Edward Esko

I hope that this book will lead to more people picking up this line of research. It is an important contribution to the development of a new paradigm of the formation of elements. — David Nagel, cold fusion researcher

In memory of

Christiane Akbar and Gene Mallove,

Who helped illuminate the way

CONTENTS

Foreword by Dr. Mahadeva Srinivasan	8
Preface: The Gentle Art of Transmutation by Alex Jack	11
Introduction by Edward Esko	42

Ideas & Experiments

1. Appearance of Silicon and Metals in Pure Graphite	56
2. Appearance of Argon in Oxygen/Helium Plasma	61
3. Appearance of Copper on a Stainless Electrode	65
4. Appearance of Palladium on a Zinc Anode	71
5. Appearance of Tin on a Silver Anode	79
6. Carbon Arc Under Vacuum	88
7. Appearance of Potassium in a Li-S Matrix	98
8. Anomalous Metals in Electrified Vacuum	103
9. In Search of the Platinum Group Metals	114
10. Quantum Tunneling and the Quantum Rabbit Effect	121
11. The Possibility of Plutonium Reduction	124
12. Lessons from Japan's Nuclear Crisis	129
13. Anomalous Metals Part II	132
14. In Search of the Platinum Group Part II	144
15. LENR-Induced Transmutation of Nuclear Waste	152
Appendix: Vanishing Metals	172
Resources	173
About the Authors	174

Periodic Table of the Elements

© www.elementsdatabase.com

- hydrogen
- alkali metals
- alkali earth metals
- transition metals
- poor metals
- nonmetals
- noble gases
- rare earth metals

1 H																	2 He
3 Li	4 Be											5 B	6 C	7 N	8 O	9 F	10 Ne
11 Na	12 Mg											13 Al	14 Si	15 P	16 S	17 Cl	18 Ar
19 K	20 Ca	21 Sc	22 Ti	23 V	24 Cr	25 Mn	26 Fe	27 Co	28 Ni	29 Cu	30 Zn	31 Ga	32 Ge	33 As	34 Se	35 Br	36 Kr
37 Rb	38 Sr	39 Y	40 Zr	41 Nb	42 Mo	43 Tc	44 Ru	45 Rh	46 Pd	47 Ag	48 Cd	49 In	50 Sn	51 Sb	52 Te	53 I	54 Xe
55 Cs	56 Ba	57 La	72 Hf	73 Ta	74 W	75 Re	76 Os	77 Ir	78 Pt	79 Au	80 Hg	81 Tl	82 Pb	83 Bi	84 Po	85 At	86 Rn
87 Fr	88 Ra	89 Ac	104 Unq	105 Unp	106 Unh	107 Uns	108 Uno	109 Une	110 Unn								

58 Ce	59 Pr	60 Nd	61 Pm	62 Sm	63 Eu	64 Gd	65 Tb	66 Dy	67 Ho	68 Er	69 Tm	70 Yb	71 Lu
90 Th	91 Pa	92 U	93 Np	94 Pu	95 Am	96 Cm	97 Bk	98 Cf	99 Es	100 Fm	101 Md	102 No	103 Lr

FOREWORD

Edward Esko and Alex Jack have continued the legacy of the pioneers George Ohsawa, Louis Kervran and Michio Kushi of the field now widely known as "Low Energy Nuclear Reactions" (LENR). I first heard about Kervran's book *Biological Transmutations* through Eugene Mallove in the early '90s and was fascinated by the concept of occurrence of nuclear reactions in living organisms. We even carried out some experiments on the possible occurrence of elemental transmutations during germination of seeds at the Bhabha Atomic Research Centre (BARC) in the mid-'90s. Although we did get what appeared to be preliminary positive results, the effort did not lead to a publication as I retired from BARC shortly thereafter. However Vladimir Vysotskii of Ukraine has during the last decade carried forward the field further, building upon the foundation laid by Kervran.

During my tenure at BARC I did have occasion to encourage some of my colleagues in the spectroscopy division to verify Ohsawa's carbon arc experiments, which were simultaneously, also replicated at Texas A & M University, since one of our colleagues, Dr. Sundaresan was doing a post doc with Prof. Bockris at that time. Both these carbon arc papers were jointly peer reviewed and published in the same issue (Nov. 1994) of *Fusion Technology*, courtesy of George Miley, its then editor, who even wrote a special editorial covering these two papers!

Much progress has taken place since those days in establishing firmly the occurrence of different types of transmutation reactions in a wide variety of LENR configurations. A review article on the experimental findings, written jointly with Edmund Storms and George Miley has appeared in the 2011 edition of the *Wiley Encyclopedia on Nuclear Energy*.

The year 2011 also witnessed great excitement in what appears to be industrial level heat production in Ni-H gas loaded LENR "reactors" wherein the inventors claim that the excess heat is caused by a nuclear reaction involving the transmutation of nickel isotopes to those of copper. However this claim has yet to be quantitatively established through reliable analysis of post run spent-Ni fuel powder samples.

During the last six years (since 2005) Edward Esko, Alex Jack, and their colleague Woodward Johnson of the Quantum Rabbit company have been quietly carrying out a series of systematic experiments with evacuated electrical discharge tubes designed to verify preselected transmutation reactions, much as what George Ohsawa did almost half a century ago (1964 to be precise). The Quantum Rabbit results were initially summarized in several articles in *Infinite Energy* magazine during the last few years and last year compiled in the form of the first edition of their monograph titled *Cool Fusion*. Their results certainly appear very promising and deserve to be independently confirmed, preferably in a proper academic setting. As conceded by the experimenters themselves, all suspicion of contamination being the cause of the apparent transmutation observations needs to be ruled out.

In this context the very bold initiative launched by Lewis Larsen of Lattice Energy LLC very recently is worth taking note of. At a time when the global price of gold is hitting the roof, Larsen has announced his interest in embarking on a semi commercial venture to produce gold using LENR devices starting from inexpensive tantalum, even inviting collaborating partners!

Lattice Energy's confidence in such an approach is based on their weak interaction theoretical model superposed on all the experimental results cited above (although other theoreticians have expressed some reservations regarding their model). I draw attention to the Larsen initiative only to highlight the farsighted vision displayed by the authors of *Cool Fusion* who had undertaken their experimental studies keeping in mind a time not too distant into the future when mankind may run out of precious heavy element metals. Their motivation in undertaking such experimental investigations was to pave the way in establishing a technology, which can be relied upon to hopefully come to the rescue of mankind when such an eventuality arises!

An added incentive is the potential application of LENR transmutation technology to convert radioactive waste left behind by the nuclear fission power industry into more benign stable elements!

Those of us who have been watching the cold fusion-LENR drama unfold from a "ring side seat" over the last couple of decades have been witness to the many twists and turns it has taken. We however feel fortunate to be part of scientific history being made. It is hoped that this second edition of *Cool Fusion* will serve as an inspiration to newcomers to the field and embolden them to embark on an exciting scientific journey, with a view to placing cool fusion low energy nuclear transmutations (cool fusion-LENT) on an even more strong scientific footing!

Good luck Edward and Alex!

Mahadeva Srinivasan
Bhabha Atomic Research Centre (BARC)
Mumbai (Retired)

27th May 2012

PREFACE
The Gentle Art of Transmutation

THE MATERIALS CRISIS

From the African savannah to the frontiers of outer space, humanity has responded creatively to ever changing climatic and atmospheric challenges, hunger and famine, conflict and war. In the process, we have scaled material and technological heights undreamed of by our ancestors. Just in the last decade, hundreds of millions of people in China, India, and other parts of Asia have exchanged a life of rural poverty or urban squalor for a middle class existence and are now shopping online, watching the World Cup on satellite TV, and driving smart carts, many equipped with GPS.

But despite globalization, rapid transportation, and instant communication, we still live in a world of alarming scarcity. Petroleum—the lifeblood of modern civilization—has peaked. The topsoil is depleted, the air is polluted, and the oceans are dying. Millions of species are on the brink of extinction. The polar icecaps and glaciers are melting, the sea levels are rising, and the pace of global warming is accelerating. One of every two people now has a cell phone—a milestone in the history of communication—but that will not lower the mercury levels in the ocean or improve the quality of the air we breathe. As a species we are at the crossroads. Either we find a new, sustainable way to live together on our wired, incredibly interconnected planet, or the age-old human conversation will end.

The dilemma is largely of our own making. Broadly, it involves the misuse or abuse of technology. As the recent earthquake and tsunami in Japan show, nuclear energy is inherently dangerous and not a viable alternative.

Instead of using the earth's resources wisely to create a just, harmonious world order, we have employed our advanced intellect and social skills to building bigger and more destructive weapons, a consumer society that functions on planned obsolescence, and a stratified economic order in which billions of people work in unhealthy environments for large, impersonal corporations that have no stake in the well being of local communities or even individual nations or states.

The dazzling splendor and affluence of modern civilization—as exhibited by Olympic extravaganzas, world expos, and space shuttle launches—is deceiving. Underlying the interrelated food, health, energy, and environmental crises is a deeper, more basic predicament. In the media it has been termed the materials crisis. We are running out of minerals—key elements or building blocks on which human culture and civilization have been constructed for tens of thousands of years. Signs of this impending collapse started to emerge more visibly in the last several years and especially over the last twelve months:

- China dramatically reduced the export of rare earth metals, a key group of elements that are instrumental in making everything from Priuses and solar panels to medical scanners and jet engines.

- The world cheered as 33 trapped copper miners were saved in a marathon rescue in Chile. But as the world's reserves of copper and other resources decline, thousands of miners continue to die every year. Whether the end result is diamond wedding rings, aluminum soft drink and beer cans, or laptops and iPods, mining is a major cause of deforestation, desertification, and other environmental destruction as well as accidental death and disability.

- In India, impoverished villagers disposing of medical waste from around the world became sick, and some died from radioactive contamination. Around the world there are millions of tons of nuclear waste in temporary facilities awaiting a permanent storage solution. Yet the demand for uranium, plutonium, cesium, and other radioactive elements for weaponry, nuclear reactors, dental and medical procedures, and medical procedures, and airport security continues to soar. (1). As the nuclear accident in Fukushima demonstrated, spent fuel rods stored at the site often pose a greater threat than the reactor itself.

- In Central Africa, civil war and strife have claimed the lives of millions of people. At the heart of the genocide is a struggle for control over mining and the lucrative trade in gems and scarce resources. The United Nations and governments, including the United States, have instituted laws prohibiting the trade in "conflict diamonds" and most recently "conflict minerals." These include tin, tantalum, tungsten, and gold from parts of the Democratic Republic of Congo and neighboring countries. These strategic metals are used by companies worldwide making medical devices, cell phones, airplanes, machine tools, and other consumer and industrial goods. (2)

- Solar, wind, and other renewable technologies offer a promising alternative to oil, natural gas, coal, and other carbon-emitting fossil fuels. However, they are not as green and clean as they appear. Their manufacture involves scarce metals and electronic components that have high social and environmental costs. China, for example, is now the worldwide leader in producing solar panels, energy efficient light bulbs, and wind turbines. But Chinese rice farmers are up in arms because the mines and high-tech plants making solar and wind components are leeching or dumping toxic waste into surrounding fields and streams, ruining crops, scarring the landscape, and poisoning the waterways. (3)

- Many of the world's leading industrial metals will run out in the next ten to twenty years. The British science journal *New Scientist* forecasts that terbium used to make the green phosphors in fluorescent light bulbs will be gone by 2012; hafnium, a crucial element in computer chips, will run out by 2017; all of the world's silver will also be gone by 2017; indium used for LCD monitors will be exhausted by 2020;

antimony, used to make flame retardant materials, will run out in 2022, and zinc will be depleted by 2037. *New Scientist* also predicted that platinum would be exhausted 2022, and "unlike with oil or diamonds, there is no synthetic alternative." (4)

- "Peak Copper Could Be Coming Soon," according to an article in the *New York Times*. The International Copper Study Group, Brook Hunt (which supplies research to the precious metals industry), and Citigroup warned that the annual supply of copper may peak at about 20 million tons in 2013. The world's 14 largest copper mines account for 40 percent of total yield and are aging, with some more than a century old. The largest, in Chile, will be exhausted by 2030. (5)

- The European Commission foresees the shortages of fourteen critical minerals. The chief new technologies contributing to the shortage are antimony tin oxide and micro capacitors for antimony; lithium-ion batteries and synthetic fuels for cobalt; thin layer photovoltaics, integrated circuits and white light emitting diodes for gallium; fiber optic cable and infrared optical technologies for germanium; displays and thin layer photovoltaics for indium; fuel cells and catalysts for platinum group elements; catalysts and seawater desalination for palladium group elements; micro capacitors and ferroalloys for niobium; permanent magnets and laser technology for neodymium and other rare earth metals; and micro capacitors and medical technology for tantalum. (6)

- In the Persian Gulf, a Japanese ship carrying zinc and lead was recently seized by Somali pirates. In Britain, theft of metals increased by 150 percent in the last two years, including 400,000 beer kegs. In Philadelphia, 2500 manhole covers and sewer grates were stolen in the past year, as scrap metal prices soared.

- Clearly, as the *Wall Street Journal* headline proclaimed, the world faces "A Metal Scare to Rival the Oil Scare." (7) Or as a report released by the National Academy of Science warned, "Virgin stocks of several metals appear inadequate to sustain the modern 'developed world' quality of life for all of Earth's people under contemporary technology." (8)

THE ALCHEMICAL QUEST

"Ruthford, this is transmutation!" Frederick Soddy exclaimed when they discovered that radioactive thorium was converting itself into radium in 1901. "For Christ's sake, Soddy, don't call it transmutation. They'll have our heads off as alchemists," replied Ernest Rutherford, the father of nuclear physics. (9)

From the earliest times, humanity has dreamed of creating a world of unlimited abundance. Whether called paradise, Eden, Arcadia, or Utopia, this vision continues to inspire and guide us as we confront these contemporary challenges.

At a practical level, the quest to create heaven on earth frequently took the form of alchemy, or transmuting mercury, lead, and other common elements into gold, silver, and other precious metals. From Thoth, the ancient Egyptian god of the arts and sciences, to Sir Isaac Newton, who formulated the mathematical laws of the universe while carrying out alchemical experiments at Cambridge University, discovering the Philosopher's Stone has captivated the most creative and daring minds.

In China, alchemy was associated with Taoism and the search for the elixir of life. However, it also had a materials aspect. In 1956, the tomb of a prominent general who died in A.D. 297 was discovered in northern China. Among its treasures were twenty ornamental metal belt fasteners.

Scientists at the Institute of Applied Physics of the Chinese Academy of Science and the Dunbai Polytechnic analyzed the fasteners and found they were made of an alloy of 5 percent manganese, 10 percent copper, and 85 percent aluminum. (10) Aluminum was not discovered until 1807 and not produced industrially until 1857. Extracting aluminum from bauxite is a complex process, involving temperatures exceeding 1000 degrees centigrade, a special oven, refraction chamber, and electrolysis. How did the ancient Chinese accomplish all of this?

An ancient Chinese tomb includes artifacts made with aluminum, which was not produced until 1857.

The Delhi Pillar is impervious to rust.

There are many anomalies like this in the historical annals. In India, near New Delhi, there is an iron column dating to the early fifth century that is impervious to rust. It, too, defies explanation. Known as the Delhi Pillar, it stands 23 feet 8 inches tall, averages 15 inches in diameter, and weighs 12,000 pounds. Its smooth, brass-like surface resists weathering and deterioration. Scientific analysis in 1911 established that it was 99.72 percent iron with trace amounts of phosphorus, carbon, silicon, nitrogen, copper, and other elements. (11) The pillar appears to have a extremely thin protective layer of blue-black oxide on its surface that, even when scraped away, reforms within several days.

Jewellery from ancient South America includes a platinum alloy produced by an unknown form of metallurgy.

In Ecuador, ornaments made out of platinum were discovered in the last century dating back to antiquity. The silvery white metal takes its name from the Spanish term, "platina del Pinto," meaning little silver of the Pinto River in Colombia. With its extreme durability and high melting temperature, platinum is one of the hardest metals to combust.

Julius Caesar Scaliger, the first European to describe it in 1557, observed that "no fire nor any Spanish artifice has yet been able to liquefy [it]." For 200 years, European scientists investigated its properties but were unable to fuse platinum with other elements.

The U.S. Bureau of Standards performed tests on the ancient Ecuadorian platinum and found that it had been combined with other unknown minerals to create an alloy with a melting point of 9000 degrees C. (12) No known ancient furnace (and few modern ones) could have produced such high temperatures.

What is the explanation for these enigmas? Could there have been an ancient scientific and spiritual civilization that mastered advance technologies that are forgotten and completely lost today?

Could transmutation have played a role in creating ancient aluminum, rust-resistant iron, platinum, and other mysterious metals? In the case of the ancient Chinese artifacts, Joseph Needham, a British biochemist, a member of the Royal Society, and author of the encyclopedic *Science and Civilization in China*, concluded that the aluminum was produced through transmutation. (13)

MODERN PIONEERS

In a more recent example, Norman Lockyer, the distinguished astronomer and discoverer of the element helium, conducted experiments in transmutation in London that defied explanation. In 1878, he reported that with a powerful voltaic current, he had volatilized copper in a glass tube, dissolved the resulting deposit in hydrocholoric acid, and formed calcium. (14) Lockyer was a member of the Royal Society and the founder of *Nature*, a leading British science magazine that continues to be published today.

A generation later, Sir William Ramsay, England's leading chemist and winner of the Nobel Prize for Chemistry, announced in 1907 that in vacuum tube experiments he had transmuted copper into lithium. (15) Ramsay's credits included discovering three important atmospheric elements: neon, krypton, and xenon. Over the next decade, Ramsay went on to conduct many more experiments and reported creating helium, neon, argon, and other gases from hydrogen, oxygen, and other fill gases and catalysts.

The pioneering experiments of Lockyer, Ramsay, and their colleagues were extremely promising. As Ramsay observed on a speaking tour of America in 1912, "The modern chemist must work for future generations. The synthetic process, really the development of comparatively recent years, is successfully solving many of the problems that are vital to the life of the people of the future. The work of the modern synthetic chemist now involves the saving of untold millions of dollars to the present and future generations."

Norman Lockyer (left) and William Ramsay (right), early experimenters in transmutation.

But World War I broke out several years later, and the breakthrough of the English scientists was eclipsed. By 1922, just six years after Ramsay passed away, the world heralded Sir Ernest Rutherford, the father of nuclear physics, as the avatar of a new age. "Way to Transmute Elements Is Found," the *New York Times*' headline proclaimed, "Dream of Scientists for a Thousand Years Achieved by Dr. Rutherford." (16)

The article described how the New Zealand-born physicist had transmuted nitrogen, sodium, aluminum, chlorine, oxygen, and carbon into helium and hydrogen.

In extolling Rutherford's achievement, Professor O. W. Richardson addressed a gathering of British mathematicians and physicists: "Rutherford has taken the direct method of bombarding the nuclei of the different atoms with the equally high-velocity helium nuclei (alpha particles) given off by radioactive substances, and examining the tracks of any other particles which may be generated as a result of the impact. . . . In some cases it appears the kinetic energy of the ejected fragments is greater than that of the bombarding particles. This means that these bombardments are able to release the energy which is stored up in the nuclei of atoms."

Following Rutherford's discovery, the world rushed headlong into the atomic age. As Europe rearmed, England, Germany, France, and Russia competed to secure the services of the best theoretical physicists to harness the destructive power of the atom. America would ultimately join them and win the race to create an atomic bomb. But a golden opportunity was missed in the early twentieth century as evidence of peaceful natural transmutation surfaced in ancient burial mounds and in modern vacuum tubes. Had humanity reflected more deeply on the choice before it, the nuclear arms race might have been avoided. Instead, we are now dealing with its legacy, including the horrors of Hiroshima and Nagasaki, the Cold War, millions of deaths from leukemia and other cancers from nuclear fallout, nuclear accidents at Three Mile Island, Chernobyl, and Fukushima, the burden of nuclear waste, the threat of nuclear terrorism, and the ever present danger of thermonuclear war through accident, miscalculation, or computer error.

Ernest Rutherford, the father of atomic energy, discovered the nucleus of the atom and pioneered in trying to harness its power.

Today we stand at a similar crossroads. In a world rapidly running out of energy and the material wherewithal to sustain itself, the choices before us boil down to Hot Fusion, Cold Fusion, and Cool Fusion. Let's look at each one.

HOT FUSION

The Manhattan Project was the prototype for all future atomic energy projects. In the name of a transcendent cause—the defeat of Hitler's Third Reich—virtually unlimited funds, the nation's best and brightest scientists and engineers, and unprecedented secrecy, misinformation, and deception were devoted to the effort. Although Germany surrendered in April 1945, before the atomic bomb was ready to use, it was dropped on Japan four months later. No sooner had World War II ended than the Cold War began.

For nearly half a century, the U.S. and U.S.S.R. and their allies engaged in a rivalry based on the doctrine of mutually assured destruction (MAD). Hundreds of atmospheric atomic and hydrogen bomb tests were conducted, thousands of ballistic missiles readied for launch, and trillions of dollars expended on both sides. The Center for Defense Information calculates that the U.S. alone spent $13.1 trillion between 1945 and 1996 on defense. (17) Both sides amassed enough firepower to destroy the world many, many times over. In a word, it was the era of overkill.

As the arms race threatened to spiral out of control, voices of moderation called for atomic energy to be redirected toward peaceful purposes. In a speech to the U.N. General Assembly in New York on December 8, 1953, President Dwight D. Eisenhower delivered an historic speech calling for nuclear energy to be used for the benefit of humanity as well as for military purposes. (18) He went on to launch an "Atoms for Peace" program to promote the peaceful uses of atomic energy. The first nuclear reactors in Iran and Pakistan were built under American auspices to supply power for civilian purposes. The ban-the-bomb movement rallied behind this cause and envisioned a future world community powered by safe, clean nuclear energy. Man's age-old dream of an end to hunger and poverty and abundance for all would be realized by beating nuclear weapons into peaceful atomic plowshares.

Until the accidents at Three Mile Island and Chernobyl few questioned the beneficial use of nuclear energy. Since then, its dangers have become well publicized. Let's review them here because nuclear energy, which has been stalled in the United States since 1973, is coming under renewed scrutiny as part of an overall strategy to prevent global warming. Many prominent environmentalists who opposed nuclear energy in the past have now endorsed it because of its low carbon footprint.

President Eisenhower delivers the "Atoms for Peace" speech at the United Nations in 1953 calling for the peaceful use of nuclear energy.

- Proliferation risks: Plutonium, the element most commonly a byproduct of the process of producing nuclear energy, can also be used for making bombs. In 2000, an estimated 620,000 pounds of civilian, weapons-usable plutonium was created. About 18 pounds are needed for one Nagasaki-size bomb. That yearly total is the equivalent to 34,000 nuclear weapons that could fall into the hands of terrorists or rogue states. (19)

- Radioactive risks: A nuclear accident can contaminate not only the local area but large parts of the world. WHO reported that fallout from Chernobyl explosion reached 1400 miles away to Scotland and exceeded 10,000 times normal. Increased cases of thyroid cancer and other malignancies, as well as mutations to plants and destruction to wildlife and the environment, are continuing. (20) Estimated cancer deaths from the Chernobyl accident range from 4000 to 40,000.

- **Economic risks:** Chernobyl cost the Soviet Union more than three times the saving accrued from nuclear power plants operated between 1954 and 1990. (21)

- **Environmental risks:** Uranium mining, refining, and enrichment and the production of plutonium produce radioactive isotopes that pollute rivers and lakes, air, land, plants, and equipment. Some of these isotopes have half-lives of hundreds of thousands of years and will remain lethal for thousands of human generations. Transporting wastes by truck and rail is extremely hazardous.

In response to such concerns, governments and the nuclear industry have been promoting *nuclear fusion power*. This is different from *nuclear fission power*. Conventional nuclear plants are based on nuclear fission, which causes artificial transmutation by exposing elements to neutrons produced by a fission, or splitting, from an artificially produced nuclear chain reaction. In contrast, nuclear fusion reactors typically call for the plasma of deuterium and tritium (isotopes of hydrogen) to be heated to temperatures of about 150 million degrees Kelvin, about 10 times hotter than the center of the sun. In gases heated to this temperature, electrons separate from their atoms, and hydrogen transmutes into helium. According to $E=mc^2$, the energy released in this transformation is potentially unlimited.

Creating such astronomical temperatures is a daunting task. But in 1997, after more than forty years of efforts around the world, scientists at the Joint European Torus (JET) at the Culham Centre for Fusion Energy in Britain achieved the high target temperature in a magnetically confined plasma. JET produced 16 MW (megawatts) of fusion power and utilized 25 MW in the process. (22)

The National Ignition Facility (NIF) at the Lawrence Livermore National Laboratory has pioneered another approach to hot fusion. The NIF is designed to focus 500 TW (terawatts) of energy onto a millimeter-scale fuel pellet with an array of 192 lasers. The fusion energy produced is expected to be 10 to 20 times greater than the light produced by the lasers. However, the facility overall is still less than 1 percent efficient.

In Southern France, the European Union, Russia, China, India, Japan, South Korea, and the U.S. began construction in 2008 of ITER (the International Thermonuclear Experimental Reactor), a magnetic fusion process known by its Russian acronym *Tokamak* utilizing a small, torus or donut-shaped reactor. It is scheduled to conduct a demonstration test in the mid-2020s in which 500MW of output from superheated plasma is produced using less than 50MW of input. The estimated cost for ITER is about $15 billion. When it is completed, it will be twice as large and sixteen times heavier than any previous fusion reactor. Each of the nine torus units will weigh about 400 tons. Altogether it will weigh 5,116 tons, measure 64 feet in diameter, and extend 37 feet in height.

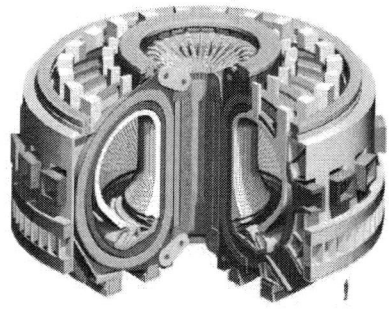

Artist's drawing of ITER, the International Thermonuclear Experimental Reactor.

On the plus side, nuclear fusion has the potential to create three times as much energy as nuclear fission and several million more times as much energy as a conventional coal reactor. It produces no CO_2 or atmospheric pollutants. Proponents say it is intrinsically safe, and there is no possibility of a catastrophic accident as with nuclear fission. Fusion energy involves small, precisely controlled temperatures, pressures, and magnetic field parameters, and damage to the reactor would cause the heat generation to rapidly cease. It utilizes radioactive materials measured in grams, while fission reactors typically store several years worth of radioactive fuel. In fission plants, moreover, heat continues to be generated through particle decay for hours or days after shutdown, leading to the possibility of melted fuel rods. A fusion reactor produces no chemical effluents, radioactive waste is substantially less, and no weapons-grade nuclear material is involved. Proponents say a fusion accident would be similar to other industrial accidents.

On the downside, hot fusion still involves major technical challenges. It requires enormous amounts of energy to initiate and control the fusion process. The plasma is relatively unstable. It produces pulsed, not continuous power, and requires a large amount of pulsed power to sustain. This could affect surrounding electrical grids. Liquid lithium is used in the transmutation process and as a coolant. It is highly flammable and could release small amounts of radioactive tritium into the environment in the event of a mishap. There are several dozen Tokamak reactors around the world, but all of them are small scale. As PhysicsWorld.Com, a pro-hot fusion site, concludes, "Despite more than 50 years of effort, today's nuclear-fusion reactors still require more power to run than they can produce." (23) Scaling up to mass production levels is challenging. After nearly fifty years of experimentation, none is economically feasible. The ITER itself is only a test facility and will not be used to produce electricity. Scientists hope that it leads to the development of actual online fusion reactors in the 2050s to 2080s.

The potential risks to the environment of nuclear fusion appear to be substantially less than nuclear fission. However, there may be serious energetic effects, especially when dealing with temperatures higher than the core of the sun. The magnetic fields in the container that encases the plasma would be more powerful than the magnetic field of the earth.

It is unclear what this vessel, the heart of the reactor, would be made of and how to prevent neutron bombardment from inducing radioactivity in the reactor itself. As French physicist Sebastian Balibar, director of research at the CNRS, said, "We say that we will put the sun into a box. The idea is pretty. The problem is, we don't know how to make the box." Hence maintaining and deactivating a commercial fusion reactor could be challenging.

It is also unknown what kind of long-term, cumulative vibrational effects hot fusion reactors would produce on workers, local communities, wildlife, and the earth as a whole. The potential hazards of cell phones, computers, and other artificial electromagnetic energy are coming under increased scrutiny. The communications industry minimizes the impact of these devices because it doesn't want to scare customers. Independent studies have linked mobile phones and similar devices with increased risk of cancer, decreased natural immunity, and other health concerns. (24)

They are also suspected of playing a role in the disappearance of the bees by disrupting their navigational systems. Compared to cell towers and satellites, hot fusion involves temperatures and magnetic fields many magnitudes greater.

Greenpeace International has campaigned against ITER saying: "Governments should not waste our money on a dangerous toy which will never deliver any useful energy. Instead they should invest in renewable energy which is abundantly available, not in 2080 but today." The activist organization says that with the money allocated to ITER, 10,000 MW offshore wind farms, delivering energy for millions of European households, could be built instead. A French network of 700 peace groups, Sortir du nucleaire (Get Out of Nuclear Energy) claims that ITER poses a health and environmental hazard because scientists do not yet known how to manipulate the high-energy deuterium and tritium hydrogen isotopes used in the fusion process.

Critics charge that nuclear fusion energy is another boondoggle like the HLC, or Large Hadron Collider, which is the world's largest high-energy particle accelerator, and the Human Genome Project, which mapped human DNA. Global projects like these are intellectually stimulating and create steady employment for thousands of scientists, but involve staggering costs and little, if any, practical results.

In a review of the Human Genome Project, the *New York Times* reported that after ten years and billions of dollars the promised benefits of identifying specific causes of disease remained unfulfilled. (25) In the *Journal of the American Medical Association*, a doctor suggested that the old-fashioned method of taking a family history was a better guide. (26)

Aside from these drawbacks, hot fusion (whether in the form of nuclear fission or nuclear fusion) deals only with the energy side of the equation, not the materials side.

In an ideal high tech world where every home, car, and community had a small portable Tokamak generating its own clean, carbon-free energy, mining would still be necessary to build the reactors and drive the global economy. A world powered by hot fusion would still be dependent on its infrastructure on extracting and refining copper, tin, iron, bauxite, gold, diamonds, and nearly one hundred other metals from the surface or deep within the earth. All of the territorial, environmental, health, and security issues related to Peak Metals would remain. Hot fusion involves transmutation, but on such a minute scale that it would have no practical impact on the supply of scarce material resources. Indeed, creating a national grid of highly energy-intensive hot fusion reactors would further accelerate mineral depletion.

COLD FUSION

According to present-day cosmology, based on the Big Bang theory, elements in the natural world cannot change into one another under ordinary conditions. Only under nuclear conditions, involving extremely high temperatures and pressures, is transmutation possible. In the fraction of a second, astrophysicists say, nearly all of the cosmos's quarks, gluons, and other elementary particles originally appeared, giving rise to hydrogen, helium, and other light elements. Since galaxies appear to be moving farther apart in space and their velocity is accelerating, scientists calculate that the primal explosion took place about 14 billion years ago at a common point of convergence. The original Big Bang theory has been modified to include the transmutation, or continuous creation, of heavier elements by the formation of supernova or by nuclear reactions such as the transformation of uranium to plutonium in a nuclear power plant or atomic bomb. Once again, these are all examples of hot fusion.

Cold fusion, in contrast, involves the transmutation of atoms and elements at conditions at much lower temperatures and pressures. These span ordinary indoor and outdoor temperatures to kitchen or tabletop science lab temperatures produced by ordinary ranges, stoves, ovens, batteries, furnaces, bunsen burners, and other low-tech devices.

Reports of cold fusion electrified the world in 1989. Martin Fleischmann, a leading British electrochemist, and Stanly Pons, an American researcher, reported that excess heat at a magnitude previously found only in nuclear processes had been created in a simple laboratory setting. (27) The experiment involved passing an

Fleishchmann and Pons electrified the world with their announcement on cold fusion in 1989.

electric current through heavy water (deuterium) on a palladium electrode in a calorimeter, an insulated chamber that measures heat. Measurements of minute byproducts of the reaction, including neutrons and tritium, suggested the creation of a low-cost, abundant source of energy.

Results of the experiment were announced in a press conference at the University of Utah and later in a scientific journal, and the process was dubbed "cold fusion." However, Fleischmann and Pons' experiment could not be consistently replicated. Some scientists failed to get any promising results, others succeeded, and still others came to mixed conclusions. The U.S. Department of Energy investigated the process and by the end of the year found that it was not persuasive. It opposed special funding for cold fusion research but supported limited funding for specific lines of further study. Despite the submission of corroborating evidence from 92 groups in 10 countries, the DOE panel noted that the reports of excess heat and the report of nuclear byproducts contradicted current scientific understanding. (28)

Within a year, the enthusiasm for cold fusion collapsed, and it became a byword for quack or pseudoscience. Cold fusion joined alchemy, psychic phenomena, ufology, and the Shakespeare authorship controversy as taboo subjects in academia.

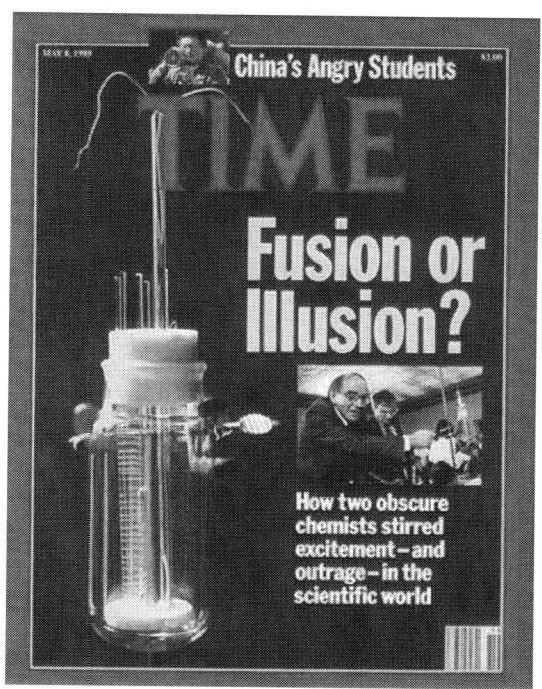

Skepticism quickly followed the introduction of Cold Fusion when it proved hard to replicate.

Fleischmann and Pons, now treated as outcasts, moved to France and received a grant from the Toyota Motor Corporation to continue their work. Their laboratory closed in 1998 after spending about $20 million in further research.

Japan's Ministry of International Trade and Industry (MITI) sponsored a "New Hydrogen Energy Program" to research cold fusion and ended research in 1997 after spending another $20 million and reporting an inability to replicate results. In 2002 researchers at the U.S. Space and Naval Warfare Systems Center in San Diego, California, who had been studying cold fusion since 1989, reported producing neutrons consistent with nuclear reactions. They issued a report "Thermal and nuclear aspects of the Pd/D^2) system," calling for further funding.

In 2008, a brainstorming meeting was organized at the National Institute of Advanced Studies in Bangalore, India, to take stock of the progress in worldwide cold fusion research. The meeting passed a resolution recommending to the government of India that serious attention should be paid to the emerging new field of LENR research.

Between 1990 and 2001, the *Journal of Fusion Technology* (FT) was established to publish peer reviewed articles on cold fusion under the editorial direction of George Miley, director of the Nuclear Laboratory at the University of Illinois, Urbana. Research articles have also appeared in several other specialized journals, including the *Journal of Electroanalytical Chemistry, Journal of Physical Chemistry, Physics Letters A, International Journal of Hydrogen Energy*, and physics, chemistry, and engineering publications in Japan, Russia, and Germany. The first International Conference on Cold Fusion (ICCF) was held in 1990, and further gatherings have been held in various countries ever since.

In a survey of ten scientific teams doing cold fusion research, a 2007 review found that the groups reported measurements of excess heat in one third of their experiments using electrolysis of heavy water or by loading deuterium gas onto lead powders under pressure. The researchers generally reported 50-200 percent excess heat generated in their tests for intervals lasting hours to days.

In a boost to the reemergence of cold fusion in recent years, *60 Minutes* featured a segment on the subject entitled "Cold Fusion Is Hot Again" on April 19, 2009. In the nationally televised CBS News program, investigative correspondent Scott Pelley interviewed scientists around the world and reported "renewed buzz among scientists that cold fusion could lead to monumental breakthroughs in energy production."

For example, researcher Michael McKubre at SRI International, a California lab that does extensive work for the U.S. government, enthused, "We can yield the power of nuclear physics on a tabletop. The potential is unlimited.

That is the most powerful energy source known to man." McKubre envisions the creation of a clean nuclear battery in 20 years that could be used for everything from laptops to cars. "The potential is for an energy source that would run your car for three, four years . . . There is ten times as much energy in a gallon of sea water, from the deuterium contained within it, that there is from a gallon of gasoline."

In Israel, Pelley visited a lab called Energetics Technologies along with Rob Duncan, vice chancellor of research at the University of Missouri and an expert in measuring energy. After reviewing cold fusion experiments for two days, Duncan reported, "I found that the work done was carefully done, and that the excess heat, as I see it now, is quite real." *60 Minutes* also obtained an internal memo from the Defense Advanced Research Projects Agency (DARPA). The Pentagon memo concluded there is "no doubt that anomalous excess heat is produced in these experiments." The U.S. military is now stepping up funding of cold fusion research at the naval research lab in Washington, D.C., and at SRI in California.

Cold fusion is very promising, as this recent revival of interest shows. Compared to hot fusion, it is simple, safe, and relatively low cost. But there are still major hurdles to be overcome before it becomes accepted by the mainstream scientific community or before it offers a practical alternative to fossil fuel or hot fusion. Results are sporadic (the success rate usually falls between 1/3 and 2/3) and difficult to replicate. Even when excess heat is produced, it can take weeks for it to appear. And the energy is never the same twice. Gathering, distributing, and delivering a uniform amount of heat or power are major challenges.

One other little recognized anomalous result of cold fusion experiments is the detection of unexpected heavy elements. Because the process focuses on the release of excess heat or energy production, it tends to ignore evidence of transmutation taking place. However, a few researchers such as Japan's Tadahiko Mizuno in Hokkaido have reported the creation of novel metals. This brings us to the third type of experiments and potential energy source: Cool Fusion.

COOL FUSION

When the future history of science is written, the name C. Louis Kervran could rank alongside that of Galileo, Newton, and Einstein. In the 1950s, Kervran, a French biochemist, was working for the French government in a project that took him to the Sahara desert. He was investigating a group of workers who were drilling wells under extremely difficult conditions. It was well known that remaining too long in the hot summer sun could deplete the body. Yet the workers toiled on unshaded metallic platforms without any apparent ill effects. For six months, a voluntary team carefully recorded everything they ingested and excreted. The results showed that as the heat increased, they eliminated more potassium through perspiration than they took in. At the same time, they consumed extra sea salt in the form of tablets they sucked while working. But the additional salt was not completely excreted. But the greatest puzzle was the thermal differential between the energy taken in and released in the form of heat. Through their diet and exposure to the sun, the workers took in an average of 4085 calories each day. At the same time, they produced 4.12 liters of perspiration a day that required 540/kilocalories per liter to evaporate. According to the classical equation, the workers should have died from hypothermia because of such a large imbalance.

"I came to the conclusion that it was sodium which, disappearing to become potassium, created an endothermal reaction (thus causing heat to be absorbed)," Kervran explained. "Hence by instinct one consumes more salt in a dry and hot country. This is why salt is so important in Africa, the Middle East, etc., where caravans traveled up to 1,000 kilometers to bring back salt." (29) Though it contradicted the basic tenets of modern chemistry, additional experiments confirmed that sodium changed into potassium. In fact, "whenever fresh sodium is injected into the organism, it is immediately transmuted into potassium." (30)

Kervran reported his findings in a scientific journal, and they came to the attention of George Ohsawa, a natural philosopher from Japan known as the father of modern macrobiotics, who immediately recognized their significance. On a visit to Paris, he subsequently met with Kervran and encouraged him to write about his experience. The result was *Biological Transmutations*, followed by *Natural Transmutations* and other works. (31)

Dr. Kervran referred to the emerging science of low energy fusion as "transmutations biologique," or "biological transmutations." Kervran, who was later nominated for the Nobel Prize in physiology, proposed that in nature, and especially within plants and animals, elements combine to form new elements through low energy fusion. The transmutation of elements occurs when two atomic nuclei fuse under low temperature, pressure, and energy. According to

Louis Kervran (left), George Ohsawa (center), and Michio Kushi (right) conducted pioneer experiments in transmutation.

Kervran, two "genesis" elements fuse under natural conditions to form a new, or "progeny" element. For example, in Kervran's hypothesis (elements shown with atomic numbers):

- Potassium (K) is formed through the low energy fusion of sodium (Na) with oxygen (O): $_{11}Na + {}_8O \rightarrow {}_{19}K$

- Silicon (Si) is formed through the low energy fusion of carbon (C) and oxygen (O): $_6C + {}_8O \rightarrow {}_{14}Si$

- Calcium (Ca) is formed through the low energy fusion of potassium (K) and hydrogen (H): $_{19}K + {}_1H \rightarrow {}_{20}Ca$

Back in Tokyo, Ohsawa determined to prove transmutation in the laboratory, but was at a loss as to how to proceed. After trying unsuccessfully to design an experiment, he went on a whole grain fast and meditated until he discovered a solution.

One night he had a dream. From the sky, among the clouds, a big hand appeared. Bolts of lightning shot forth from the fingers. They descended to earth with great brightness. As each bolt struck the ground, it exploded, leaving newly created elements behind. When Ohsawa awoke from the dream, he knew he had found the answer: electricity would be needed to make transmutation happen.

Ohsawa contracted a friend who was a professor at a large university and sought his assistance in setting up the experiment. In a very simple procedure (see Figure 1), they took a vacuum tube and attached electrodes at either end: one positive, the other negative. They attached a wire to the electrodes. They put 2.3 mg of sodium inside the tube, at the middle of which they placed a bulb containing oxygen. They ran an electric charge through the wire, which, after some time, caused the sodium to melt into a liquid, gaseous, and finally a plasma state. They placed a prism in front of the tube, and an orange color appeared, a reflection of the wavelength emitted by the sodium. At that point, they opened a valve, allowing the oxygen to enter the tube. Reasoning on the basis of traditional Far Eastern philosophy that since sodium is yang (or hard and condensed) in comparison to oxygen, which is yin (or soft and expanded), nuclear fusion between the two elements took place. They did not unite side by side as they do in a compound. Instead their atoms fused and gave rise to new atoms. At that moment, the color coming through the prism changed from orange to purple, or from yang to yin. The result of the test was that 2.3 mg of sodium fused with 1.6 mg of oxygen and formed 3.9 mg of potassium.

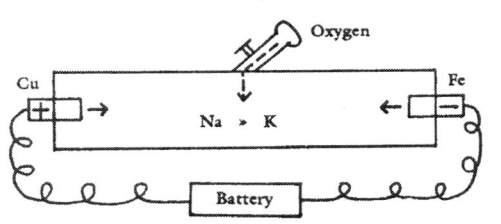

Figure 1. Ohsawa's initial experiment

This experiment helped explain how the workers Kervran studied were able to change sodium into potassium within their bodies. The workers were taking sodium in the form of salt. Oxygen is abundant in the air. And if they were working hard and breathing heavily, there would be even more oxygen in their bodies. They were yang, active people, doing yang, strenuous work in a yang, hot environment. Their bodies were highly charged with electromagnetic energy, similar to the electric charge passed through the vacuum tube. One of the fundamental principles of nature is that at its extreme yin changes to yang, and vice versa. The combination of factors caused sodium (yang) to change into potassium (yin) in their bodies. The Ohsawa team went on to conduct several other breakthrough experiments, producing iron from carbon, and tiny amounts of several precious metals, including gold and platinum. (32)

THE SPIRAL OF THE ELEMENTS

The spiral arrangement of major elements is made according to the spectroscopic examination of color waves. (33) Approximately 8,000 Å (angstroms) to 5,000 Å is the area of yang elements and approximately 5,000 Å to 3,500 Å of yin elements. According to this chart, elements occupying the positions in opposite orbits — e.g., hydrogen (H) and oxygen (O) can combine easily due to the principle of attraction between yin and yang; and elements occupying a similar position — e.g., hydrogen (H) and helium (He) — have difficulty combining with each other unless technical changes of

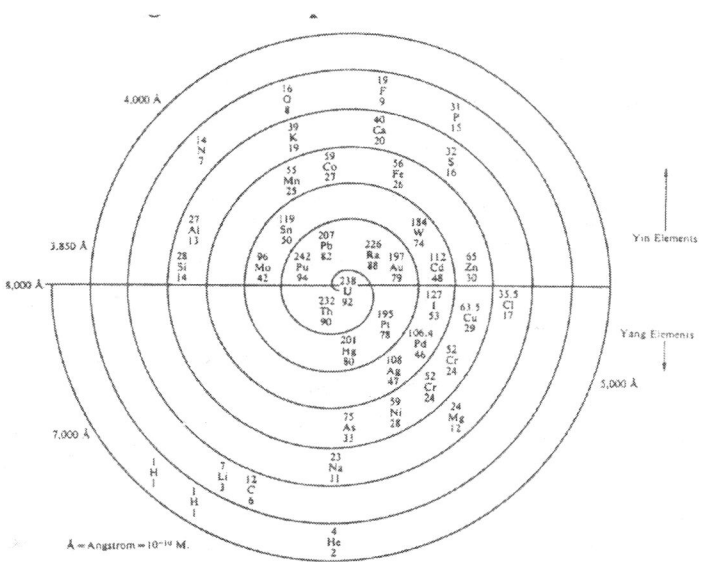

Figure 2. The Spiral of Elements © 1994 by Michio Kushi

temperature, pressure, or nature are applied.

Elements occupying peripheral areas are more yin and lighter—e.g., hydrogen (H) and helium (He)—while elements located at more central areas are more yang and heavier—e.g., zinc (Zn) and iron (Fe). Elements belonging in the most central orbits—e.g., uranium (U) are radioactive, tending to return to the outer orbits in the same way the sun is radiating its energy outwards in the solar system. Most balanced elements are found in the fourth orbit, and some of them are magnetic, such as iron (Fe), cobalt (Co), and nickel (Ni).

The chart reveals that lighter elements are gradually transmuting towards heavier elements and heavier elements are in turn transmuting back into lighter elements, though it may take some thousands to millions of years to do so naturally. The transmuting speed of peripheral elements is much slower than that of central elements.

The precise chart of classification of the elements by yin and yang should be considered together with this spectroscopic examination, including other factors such as the nature of chemical reactions and freezing, melting, and boiling temperatures. Knowing the yin and yang natures of elements helps us understand all phenomena—chemical and biochemical, geological and biological—as well as the order of change. Michio Kushi (1926-), who had studied with Ohsawa and went on to introduce macrobiotics to America, began to see how the transmutation of elements directly reflected the spiral formation of the universe through seven stages, from infinity to the world of matter.

Together with Ohsawa, he constructed a comprehensive spiral chart of the elements, arranging them from yin to yang. This chart is more practical, dynamic, and realistic than the current periodic table. The elements are actually stages in a continuum. They are not separate and distinct. They do not exist independently of one another, but are part of a spiral process of change. Even though it may take billions of years, a hydrogen atom will eventually change into heavier and heavier elements. And once it reaches the heaviest (radioactive) state, it begins to change back to the lightest stage—the hydrogen atom. The movement from expansion to contraction and from contraction back to expansion occurs everywhere in nature. The behavior of atoms and elements is but a reflection of the endless order of change. Kushi's collected writings on transmutation were subsequently published in *The Philosopher's Stone*. Over the last generation, researchers at the University of Texas, at the Bhabha Atomic Research Centre in India, and elsewhere have replicated some of the seminal experiments conducted by Kervran, Ohsawa, and Kushi. During the Cold War, American military scientists tested the theory of biological transmutation and verified the peaceful transmutation of matter from cell to cell and atom to atom. The U.S. Army study, subsequently declassified, even proposed a possible biological mechanism for the process:

> The works of Kervran, Komaki, and others were surveyed, and it was concluded that, granted the existence of such transmutations (Na to Mg, K to Ca, and Mn to Fe), then a net surplus of energy was also produced. A proposed mechanism was described in which Mg adenosine triphosphate, located in the mitochondrion of the cell, played a double role as an energy producer. In addition to the widely accepted biochemical role of MgATP in which it produces energy as it disintegrates part by part, MgATP can also be considered to be a cyclotron on a molecular scale. The MgATP when placed in layers one atop the other has all the attributes of a cyclotron in accordance with the requirements set forth by E. O. Lawrence, inventor of the cyclotron. It was concluded that elemental transmutations were indeed occurring in life organisms and were probably accompanied by a net energy gain ....
>
> The relatively available huge supplies of the elements which have been reported to have been transmuted and the probable large accompanying energy surplus indicate a new source of energy may be in the offing—one whose supply would be unlimited. (34)

In the world of chemistry, the discovery of transmutation paralleled the revolutionary insights that transformed the other sciences. Galileo, Newton, and the seventeenth century natural philosophers proved that the earth was not the fixed center of the cosmos, but a small planet in a constantly changing universe. In the nineteenth century, Darwin broke with millennia of dogma and showed that species were not immutable but evolved, or descended, from common ancestors and would give rise to new forms of life in the future.

The modern belief that chemical elements are static and cannot be transmuted except by atomic fission, hydrogen fusion, or other high tech methods has led to the development of the Manhattan Project, the Supercollider Project, and other complex, prohibitively expensive campaigns to unlock the secrets of the atom. A similarly misguided and wasteful effort has been expended on the Human Genome Project based on the assumption that genes are fixed and constant and trigger specific diseases and disorders. In actuality,

The Bhabha Atomic Research Centre in India has replicated some of the Ohsawa experiments.

genes are dynamic and interactive, part of a complex web of physical, mental, spiritual, and environmental factors that influence and shape our health and destiny.

In contrast to these complex, destructive approaches, simple, peaceful methods using common, ordinary elements and compounds promise an abundant source of cheap, renewable energy. Cold fusion appears to be a special case of transmutation. The introduction of cold (an extreme yin factor) produces contraction (a strong yang factor) that causes certain elements to change mass, weight, or charge as electrons and other particles condense at extremely low temperatures, form new combinations, and release excess heat.

The discovery of the peaceful, natural transmutation of elements offers an alternative to destructive, artificial methods of energy production. Atomic transmutation is taking place in nature and can be duplicated without having to attack and destroy atoms. The work of Lockyer, Ramsay, Kervran, Ohsawa, Kushi, and other pioneers in transmutation has shown that matter is not fixed and static, but dynamic and changing. In principle, all scarce, precious elements can be produced from safe, commonly available elements. In *Cool Fusion*, we present the most recent chapter in this story. Quantum Rabbit was founded in 2005 to continue the work of the 19th and 20th century pioneers and help solve the problems of the 21st century.

This new science has the potential to eclipse the discovery and use of nuclear power, revolutionize science and technology, and open the door to a new era for humanity. If this knowledge is properly understood and applied, it will contribute to an era of peace and unprecedented prosperity. Now that the basic theory has been proved, the task that remains is how to develop it practically and on a large scale. In addition to creating a world of material abundance for everyone, we now have the ability to clean up the old one. Through quantum conversion, all harmful, toxic elements can be transmuted into safer, more stable ones. This includes heavy metals used in agriculture and industry that have polluted the environment. Radioactive elements, including iodine 131, cesium 137, strontium 90, plutonium 239, and uranium 235 can also be rendered harmless. Their half-lives range from years and centuries to millennia and eons.

The principles introduced in this book offer a simple, practical solution to this dilemma. The peaceful transmutation of the atom—the dream of Einstein, Oppenheimer, and the original scientists who developed the atomic bomb; to presidents Eisenhower and Kennedy; and to all who lived during the Cold War—may hold the key to the transmutation of society itself and the preservation of the planet.

Alex Jack
Becket, Massachusetts
May 19, 2012

Notes:

1. Nadene Ghouri, "Medical Waste Scandal Scars Gujarat," BBC News, March 6, 2009.
2. Chelsey Drysdale, "U.S. Enacts 'Conflict Metal' Law," *Circuits Assembly*, July 21, 2010.
3. Keith Bradsher, "Earth-Friendly Elements, Minded Destructively," *New York Times*, December 26, 2009.
4. David Cohen, "Earth's Natural Wealth: Audit," *New Scientist*, May 23, 2007.
5. *New York Times*, September 2, 2009.
6. Philip Burgert, "EU Foresees Shortages of 14 Critical Minerals," www.resourceinvestor.com.
7. "A Metal Scare to Rival the Oil Scare," *Wall Street Journal*, May 25, 2007.
8. NAS study
9. "Alchemy, Long Scoffed At, Turns Out To Be True," *New York Times Magazine*, February 19, 1911.
10. Joseph Robert Jochmans, Lit.D., "Secrets of the Alchemists: Is Modern Science Simply Rediscovering Lost Ancient Knowledge? Aluminum in Ancient Times," *Atlantis Rising*, Nov./Dec. 2008.
11. Joseph Robert Jochmans, Lit.D., "India's Magical Ancient Pillar: The Delhi Pillar Is a Genuine Out-of-Place Artifact," *Atlantis Rising*, May/June 2009.
12. "Sophisticated Metallurgical Skills in Ancients," June 13, 2010, www.maya12-21-12 on: June 13, 2010
13. Quoted in Jochmans, op cit, 2009.
14. "Transmutation of Metals: Is the Old Dream of the Alchemists to Be Realized?" *New York Times*, December 13, 1878.
15. "Turns Copper into Lithium: Sir William Ramsey Effects the Tran smutation of Elements, Sought by the Alchemists," *New York Times*, January 27, 1907.
16. "Sir William Ramsay, Noted Chemist, Dies," *New York Times*, July 24, 1918 and "Way to Transmute Elements Is Found: Dream of Scientists for a Thousand Years Achieved by Dr. Rutherford," *New York Times*, January 8, 1922.
17. Martin Calhoun, Senior Research Analyst. "U.S. Military Spending, 1945-1996 ," Center for Defense Information, July 9, 1996.
18. President Eisenhower's "Atoms for Peace" Speech, December 8, 1953, www.atomicarchive.com.
19. Leslie Lai & Kristen Morrison, "Nuclear Energy Fact Sheet," Nuclear Age Peace Foundation, 1998-2011.
20. "Nuclear Radiation and Health Effects," World Nuclear Association, June, 2010.
21. "Nuclear Plant Risk Studies: Failing the Grade," Union of Concerned Scientists, 2010.
22. "Joint European Torus," Wikipedia.org.
23. "Hot Fusion," December 6, 2010, PhysicsWorld.com.
24. Devra Davis, Disconnect: The Truth about Cell Phone Radiation," Dutton, 2010.
25. Nicholas Wade, "A Decade Later, Genetic Map Yields Few New Cures," *New York Times*, June 12, 2010.
26. Ibid.
27. "Cold Fusion," Wikipedia.org.

28. Ibid.
29. C. Louis Kervran, *Biological Transmutations*, Happiness Press, Asheville, N.C., 1980, pp. 27-28.
30. Ibid., 29.
31. Kervran published books and articles on low energy fusion in French, including *Transmutations Biologiques*, Librairie Maloine S.A., Paris, 1968; *Transmutations Naturelles, Non Radioactives*, LibrairieMaloine S.A., Paris, 1963; and *Preuves Relatives a L'Existence de Transmutations Biologiques*, Librairie Maloine S.A., Paris, 1968. Several of his works have been published in English as *Biological Transmutations*, Happiness Press., Asheville, NC, 1980.
32. Michio Kushi with Edward Esko, *The Philosopher's Stone*. One Peaceful World Press, Becket, Mass., 1994.
33. The Spiral of the Elements chart and sidebar are reprinted from Michio Kushi and Alex Jack, *The Book of Macrobiotics*, Tokyo and New York: Japan Publications, 1984, pp. 42-42. Reprinted with permission.
34. Solomon Goldfein, "Energy Development from Elemental Transmutations in Biological Systems," Report 2247 (Ft. Belvoir, Va.: U.S. Army Mobility Equipment Research and Development Command, 1978).

INTRODUCTION

Edward Esko speaking at the 2009 Conference on Future Energy. George Miley at right.

On the drive back from the Third Conference on Future Energy in Washington, DC, Florence Johnson hands me the latest issue of *Macrobiotics Today*. She points to the article, "George Ohsawa's Last Letter—To his students in the United States." It is October 2009 and, with Woody Johnson at the wheel of his hybrid SUV, we are heading back to Massachusetts after making a presentation at the Washington DC Conference. It is a fine sunny day along the Eastern Seaboard.

In his "Last Letter," Ohsawa states, "On New Years Day, 1964, I began the final course of Lao Tsu: alchemy. Surprisingly, I finished it very rapidly, for as soon as I had gathered the equipment necessary for experimentation, I achieved the transmutation of sodium to potassium (Na to K). The date was June 21, 1964 — only five months later.

"Upon my return from a world trip, I began experimenting with the transmutation of carbon into iron (C into Fe). This was also successful. Is it not wonderful that when I asked that sodium change to potassium, it happened? And when I asked that carbon change to iron, it was no sooner said than done?

"By November I was able to conclude that all elements up to atomic number 82 (lead) could be transmuted from lighter elements like carbon, oxygen, or lithium. This achievement is proof of what can be realized through a deep understanding of one of the fundamentals of the macrobiotic principle: Nothing is eternal — everything changes in this relative world."

Ohsawa died a year after writing this letter. He didn't have the chance to develop these discoveries beyond the initial stage. Had he had another ten years to develop his transmutation work, the world we inhabit today would be vastly different. We would be much nearer to a sustainable future based on universal health and prosperity.

For me, reading Ohsawa's letter in 2009 could not have been timelier. I had just given a PowerPoint presentation in which I described five low energy nuclear reactions (transmutations) achieved at our Quantum Rabbit lab. Quantum Rabbit is in essence an advanced study group that I started together with Alex Jack and Woody Johnson. Its purpose is to explore and develop new applications of yin and yang (yin/yang apps) in a variety of domains, beginning with the world of elements and energy.

After more than thirty years studying how to use the vegetable world in diet, health, and healing, including the health of society and the planet as a whole, I decided it was time to move on to the next level of study: the world of elements. Studies began in earnest early in the year 2000, even though I had lectured on Ohsawa's research beginning in 1975 at the East West Foundation and later at the Kushi Institute. In 1994 I edited Michio Kushi's lectures on this topic and added original insights. One Peaceful World Press published this as a small book, *The Philosopher's Stone*.

As Michio Kushi once said, after you reach the top of the mountain, you look out and see an even higher mountain in the distance. That higher mountain is the seemingly fixed and impenetrable barrier known as the elemental world. Unlike the soft, feminine world of plants, which readily yield to the cut of the knife and fire of the stove, the solidly masculine world of elements appears as hard and unchanging as a stone on the beach, the imposing rock faces on Mt. Rushmore, and the brilliant 24-carat diamond in an engagement ring. As the saying goes, "diamonds are forever."

Although it is acknowledged that atoms are composed primarily of empty space, with no solid substance, the periodic table gives no indication that elements can in any way change into one another. On the surface at least, the elements seem to occupy a fixed position in a universe that is static and unchanging.

It was this world that we decided to challenge. It was this barrier that we sought to breach so as to enter a new, quantum reality unbound by artificial constructs and governed only by the endless law of change. We would work and become familiar with the chemical elements just as we had previously worked and became familiar with brown rice, carrots, barley, sea salt, and cabbage in our kitchen workshops. This time, our workshop would take the form of a small laboratory with vacuum equipment, an electric power supply, and pure element samples.

The origins of alchemy are shrouded in myth and legend. Ancient Egyptian writings describe "visitors from the firmament" who shared their knowledge of the universe with humanity, including the practice of alchemy.

Supposedly, godlike beings arrived in Egypt who possessed an advanced spiritual technology through which they were able to transform matter. Medieval alchemists are the forefathers of modern chemists. Alchemy was practiced in ancient Greece, Arabia, Europe, India, and China, in addition to its legendary origins in Egypt. More recently, we trace our lineage to Sir Norman Lockyer, founder of the British publication *Nature* and discoverer of the element helium. An article published in the *New York Times* on December 13, 1878 describes Sir Norman's exploits. The article, titled "Transmutation of Metals—Is the Old Dream of the Alchemists to Be Realized?" talks about his curious results:

A correspondent from the *London Daily News* writes: "Today, in the presence of a small party of scientific men, Mr. Lockyer, by the aid of a powerful voltaic current, volatized copper within a glass tube, dissolved the deposit formed within the tube in hydrochloric acid, and then showed by means of the spectroscope, that the solution contained no longer copper, but another metal, calcium, the base of ordinary lime. The experiment was repeated with other metals and with corresponding results. Nickel was thus changed into cobalt, and calcium into strontium. All these bodies, as is well known, have ever been regarded as elementary, that is, incapable of being resolved into any components, or being changed into one another. It is on this basis that all modern chemistry is founded, and should Mr. Lockyer's discovery bear the test of further trial, our entire system of chemistry will require revision. The future possibilities of the discoveries it is difficult to limit...."

"Mr. Lockyer is one of our best living spectroscopists, and no man with a reputation such as his would risk the publication of so startling a fact as he has just announced to the scientific world without the very surest grounds. He was supported yesterday by some of our leading chemists, all of whom admitted that the results of his experiments were inexplicable on any other grounds but those admitting of the change of one element into another..."

Apparently, Sir Norman achieved his transmutations using a very basic method, without super-high temperatures, pressures, or energy. Unfortunately, nothing seems to have come from these experiments. Meanwhile science was busy pursuing another approach to transmutation, an approach with potentially sinister and destructive implications: the bombardment of atoms with radioactive particles. In 1907, Sir William Ramsay, one of England's top chemists, announced that he had achieved the transmutation of elements using the radioactive "emanations" of radium. His announcement was published in the *New York Times*, July 27, 1907, in an article titled, "Turns Copper into Lithium," with the subtitles: *Sir William Ramsay Effects the Transmutation of the Elements Sought by the Alchemists—Change Caused by Radium, Theory that the Elements of High Atomic Weight will Disappear—Statement by Ramsay."*

LONDON, July 26—Sir William Ramsay has promised to communicate shortly to the Chemical Society an account of a discovery, which, in the words of so conservative a scientific publication as *The Lancet* in its number issued today "marks an epoch in the history of chemical science, since his investigations have shown that a given element under the powerful action of radium emanations undergoes degradation into another. In short," adds *The Lancet*, "the transmutation of the elements is actually un, fait accompli." Reversing the process sought by ancient alchemists, who believed that there was a substance by means of which the baser metals could be transmuted into the higher, Sir William has effected the degradation of metals by means of gas evolved from radium. The paper will prove that Sir William has degraded copper to the first member of its family, namely, lithium. In other words, he has effected the transmutation of copper. Commenting on Sir William's experiments, *The Lancet* continues: "These remarkable discoveries remind us again of the extraordinary prescience of the ancients and of the presentiments of the alchemists, who evidently had some sort of conviction that after all there is a primary matter from which all other elements are formed by various condensations. He is a bold man who nowadays confesses skepticism about anything. The world has seen men who have said 'it is impossible,' and generations who succeeded them who have seen the impossible come true."

In the thirty years between Sir Norman's remarkable experiments and Sir William's demonstrations, the discovery of radioactivity by Marie Curie in 1897 would prove decisive, especially the radioactive properties of uranium. It was these radioactive "emanations," specifically those of radium that enabled Sir William to achieve his novel results.

From then on, alchemy took what I consider to be a decidedly wrong turn, transforming itself from what I call "golden" alchemy, or the pursuit of transmutation using relatively natural (low energy, temperature, and pressure) conditions for the benefit of humanity, to what I term "dark" alchemy.

Dark alchemy utilizes dangerous radioactive elements to produce nuclear reactions (fission and fusion) through the deployment of highly destructive, expensive, dangerous, and wasteful technologies with potentially catastrophic consequences.

The discovery of the radioactive properties of uranium, one of the extremely heavy elements at the far end of the periodic spectrum, was the driving force behind these developments. Today's arsenals of atomic and hydrogen bombs are the legacy of this latter approach, as is the proliferation of atomic weapons and ever-increasing stockpiles of plutonium and other highly toxic materials.

It was in the midst of the nuclear nightmare spawned by 20th century physics that the first rays of purifying sunlight began to appear in the person of Louis Kervran.

Kervran appeared out of nowhere like the proverbial lotus flowers emerging from the proverbial mud. Kervran was the tiny dot of Yin within the larger Yang; the tiny Yang within the all-encompassing Yin. Although small and seemingly insignificant, the contrary dots contain the seeds of great change. They are harbingers of things to come, like the tiny mammals that scurried to avoid being trampled by the giant dinosaurs.

Through his discovery of biological transmutations, Kervran predicted 21st century physics and technology, in which science would accept and develop golden alchemy, triggering a paradigm shift capable of averting further human and ecological catastrophe. Through years of patient observation, Kervran, a biochemist from Brittany in France, discovered that transmutation occurs naturally, at no cost, and quite peacefully in the life cycle of plants and animals. Kervran's formulas were an organic, refreshing, and much needed antidote to cyclotrons, aboveground hydrogen bomb tests, and nuclear reactors. It had finally come down to a choice between organic alfalfa sprouts or reactor waste.

Kervran's well-known formula, $_{11}Na_{23} + {_8}O_{16} \rightarrow {_{19}}K_{39}$ (sodium-23 + oxygen-16 into potassium-39), a process he observed in the human body, would become the cornerstone of Ohsawa's experiments. It would also serve as the basis for Quantum Rabbit low energy formulas, including several in which a highly toxic radioactive element such as plutonium can be fractioned into a benign non-radioactive element such as bismuth or lead.

When I spoke in August 2008 at the Transmutation Workshop organized by Prof. George Miley (a nuclear fusion expert from the University of Illinois) at the International Conference on Cold Fusion (ICCF-14) held in Washington, DC, I asked the group of about fifty international scientists if they knew of Louis Kervran. Practically everyone raised his or her hand. When asked, a surprising number also knew about Ohsawa's transmutation work.

Edward Esko speaking at the 2008 ICCF Transmutation Workshop. George Miley at right.

Aside from the mini-revival of carbon-arc studies in the 1990s at Texas A & M University, in India, Japan, and by Michio Kushi and Chris Akbar in Brookline, inspired by Ohsawa (and by publication of *The Philosopher's Stone*); and *all* of which confirmed Ohsawa's earlier findings, I had assumed that research on low energy transmutation had come to a standstill due to inertia and indifference on the part of the scientific community. As it turned out, I was wrong.

The first indication of this came when Alex, Woody, and I went to the Massachusetts Institute of Technology (MIT) in September 2007 to present the results of our carbon-arc experiments.

Armed with PowerPoint slides, video clips, lab reports, and formulas we entered the citadel of modern chemistry and physics to meet with MIT Professor Dr. Peter Hagelstein, one of a handful of mainstream scientists who have continued with research on cold fusion following the highly controversial 1989 disclosure of this potential source of free energy.

Although the focus of cold fusion is on the creation of energy, a number of investigators have noted the creation of new elements during the process, adding possible confirmation to the theory of low energy transmutation. Dr. Hagelstein listened to our presentation and offered valuable suggestions, including adding high-energy particle detection slides to our studies, to see if the transmutation process produces energy.

Interestingly, in our carbon-arc studies, based on those of Ohsawa, we discovered not only the possibility of transmuting carbon into iron (with oxygen from the atmosphere), but also a host of other reactions involving *both* carbon (yang) and oxygen (yin) and carbon (yang) and nitrogen (yin) from the surrounding air.

Our pure graphite (carbon) test samples showed, in addition to iron, the presence of magnesium, aluminum, silicon, scandium, titanium, cobalt, and nickel—in what appeared to be a veritable cascade of transmutations.

We reported these to Dr. Hagelstein. We also showed clips in which we tested graphite powder for magnetic properties. Using powerful magnets composed of neodymium, one of the rare earth metals, all of the treated samples showed magnetic activity, while untreated samples did not.

Dr. Hagelstein told us of other work being conducted on cold fusion and transmutation around the world. He mentioned the work done at Mitsubishi Heavy Industries in Japan in which researchers reported the ability to transmute one heavy element into another, notably cesium (Cs) into praseodymium (Pr) and strontium (Sr) into molybdenum (Mo). Fifteen laboratories in six countries have reported evidence for transmutation. Isotope ratios that deviate from natural abundances have also been reported.

Such variation in the isotope distribution of an element is considered evidence it may have formed through transmutation rather than being present beforehand as a contaminant.

Alex Jack (left) and Woody Johnson at MIT.

Our vacuum studies have been conducted at small labs in Nashua, New Hampshire and Owls Head, Maine at Rockland Harbor. We had to more or less start from scratch, reinventing the wheel so to speak, as details of the Ohsawa experiments were sketchy at best. We were entering uncharted territory, out on the open sea with only the compass of the unifying principle to guide our direction.

Fortunately, we teamed up with one of the top vacuum consultants on the East Coast, along with a master scientific glassblower who fabricated beautiful and functional vacuum tubes according to my design specifications.

We started initially with experiments on the noble gases, helium (He), neon (Ne), argon (Ar), and krypton (Kr). After much trial and error, we achieved what appeared to be the transmutation of helium (He) into argon (Ar), with the addition of oxygen (O2), with the applied formula: $2He_4 + 2(_8O_{16}) \rightarrow {_{18}}Ar_{36}$.

This formula seemed to confirm Ohsawa's hypothesis that a yang element (helium) would readily fuse with a yin element (oxygen) to form a new, hybrid element, argon.

From the noble gases we moved on to research with metals, including several attempts to duplicate Ohsawa's sodium into potassium experiment. We failed repeatedly. Undaunted, we persevered in our studies. It was not until May of 2008 that we stumbled upon an unexpected roadmap to guide us in our research. The key was again the relationship between sodium and potassium. If you look at the periodic table, you see that the element sodium (atomic number 11) and potassium (atomic number 19) are separated by an element with the atomic number 8. That element turns out to be oxygen ($_8O_{16}$). In other words, it is an atom of oxygen that separates sodium from potassium, as Kervran discovered and Ohsawa demonstrated.

Looking further at the periodic table, we see that sodium and lithium have a similar relationship. In other words, lithium, with the atomic number 3 and sodium, with atomic number 11, are also separated by atomic number 8; once again, an atom of oxygen.

These three metals, lithium, sodium, and potassium are related; they share many characteristics as members of the Group IA elements known as the Alkali Metals, for example low melting temperatures, light weight, softness (you can cut them with a knife), and extreme volatility. Sodium, for example, will explode if placed in water.

Using Ohsawa's principle of change, we see that lithium is actually the precursor to sodium, and sodium the precursor to potassium, each separated by oxygen. With this background I designed an experiment in which we would try to *quantum bounce* lithium further along the periodic table, so that it changed into sodium and then to potassium, according to the applied formula: $_3Li_7 + {_8O_{16}} \rightarrow {_{11}Na_{23}}$. $_{11}Na_{23} + {_8O_{16}} \rightarrow {_{19}K_{39}}$ (lithium-7 + oxygen-16 $\rightarrow$ sodium-23; newly formed sodium-23 + oxygen-16 $\rightarrow$ potassium-39.)

Another variation of the reaction is: $_3Li_7 + 2({_8O_{16}}) \rightarrow {_{19}K_{39}}$, or lithium-7 fuses with two atoms of oxygen-16 (a molecule of oxygen, or O_2), skipping sodium-23 and *quantum bouncing* to potassium-39.

I devised two experiments to test this hypothesis. In the first two, conducted on February 29 and May 2, 2008, we used pure lithium metal, stainless electrodes, and pure oxygen as fill gas in the vacuum tube.

All the test samples showed the presence of sodium (at up to 0.94%) and potassium (at up to 0.14%.) We were excited, thinking we had proven the formula. I wrote a letter to *Macrobiotics Today* announcing this result. However, upon closer examination, we discovered that the borosilicate (Pyrex) glass used in the vacuum tube contained a significant trace of sodium and minor trace of potassium, enough to influence the outcome of the experiment. I scheduled another series of tests for May 30 with the goal being to control for sodium. A new tube was designed so that the reaction zone at the center would be made of quartz. The quartz contained less than a part per million sodium, so we were confident it wouldn't influence the outcome.

After conducting the test, we noticed that the tips of the stainless steel electrodes had undergone a color change, with a copper-colored residue on the surface.

We sent the materials off to the outside lab for third-party analysis. When the results came back, we were initially disappointed. There were no detectable traces of sodium or potassium, although copper was present, for no apparent reason, on one of the stainless electrodes. I remember Alex commenting somewhat dejectedly, "There's nothing there." It took several days for us to realize that we had achieved an unexpected result: the presence of a significant trace of copper on the surface of the stainless electrode. It suddenly dawned on me what we had done. We had apparently caused lithium (Li-7) to fuse with the iron (Fe-56) in the stainless steel electrode to produce copper (Cu-63), according to the applied formula: $3Li_7 + {}_{26}Fe_{56} \rightarrow {}_{29}Cu_{63}$.

I realized that we could now use metallic lithium (3), the first solid element, following hydrogen (1) and helium (2) on the periodic table, as the catalyst or trigger for a wide range of low energy transmutations.

Interestingly, when lithium metal is vaporized in the vacuum tube, it emits a deep ruby red glow, which is apparently yang. Classical alchemists described the Philosopher's Stone, the mysterious substance used to transmute the elements, as having a ruby red color. Coincidence?

I set out to test the lithium hypothesis in several additional tests. In one, I predicted that lithium-7 would fuse with silver-109 to form tin-116. In another, that copper-63 would fuse with lithium-7 to produce germanium-70.

These formulas were confirmed in the autumn of 2008. As predicted, tin (atomic number 50) appeared on the surface of the silver anode (atomic number 47) following a low energy nuclear reaction with lithium (atomic number 3). Germanium (Ge) has the atomic number 32. It has repeatedly appeared in tests in which copper electrodes (atomic number 29) were used with lithium metal (atomic number 3).

Our success with lithium prompted me to search for another solid catalyst. Sulfur (S) turned out to be the perfect candidate. In one test, we were able to fuse both lithium and sodium with sulfur to produce potassium, a variation of Ohsawa's sodium to potassium experiment.

We were able to produce potassium-39 from the fusion of lithium-7 and sulfur-32. In addition, sulfur may have fissioned into two atoms of oxygen-16, each of which fused with an atom of sodium -23 to form potassium-39. In our view, sulfur-32 is formed by the fusion of two atoms of oxygen-16. Sulfur is actually crystallized oxygen.

In another test, we used a zinc (atomic number 30) electrode in combination with sulfur (atomic number 16) and oxygen test material. As predicted, palladium (atomic number 46), a rare member of the platinum group of metals, was found after the test. I had predicted this with the formula: zinc-68 + sulfur-34 $\rightarrow$ palladium-102.

Many discoveries lie ahead. Our work so far on transmutation has been unsophisticated and tentative, kindergarten at best. Whether these results lead to a Golden Age of peace and prosperity remains to be seen. However, we are confident that our studies, together with those of our colleagues around the world, shall establish once and for all Ohsawa's most fundamental principle: "Nothing is eternal. Everything [and most certainly the chemical elements] changes in this relative world." That much is certain.

Ohsawa concluded his letter by stating, "In closing, let me sincerely urge you to study even more deeply than ever before."

I can answer Ohsawa by stating: "George we will try our best. We know that the unifying principle, which you taught for more than fifty years, holds the key to health, peace, and happiness on this earth. In our time we will do our best to develop the unifying principle in the areas we deem most vital to human health and happiness. It is this dream we hope to pass on to future generations."

It is my hope that this small book will help move us in that direction.

Edward Esko
Pittsfield and Lenox, Massachusetts
June 2011

Notes on the Revised Edition

Things change rapidly in the fields of new science and new energy. Since publication of *Cool Fusion* in the summer of 2011, the QR team met with Mahadeva Srinivasan, former director of the physics group at the Bhabha Atomic Research Centre in India. Dr. Srinivasan joined Alex Jack, Michio Kushi, and me in Boston to discuss the potential of utilizing LENR-induced transmutation as a means for reducing nuclear waste and solving other problems of ecology, economy, and natural resources.

In this new updated edition, I have included two articles on QR research conducted on Sept. 27, 2011. In the first article, I report on a QR experiment on the possible reduction of lead-204 and the anomalous appearance of gold-197. This experiment duplicates earlier QR research on the possibility of initiating the low energy fission of heavy elements through a low energy fusion reaction, or cool fusion → cool fission. I have also included a paper on the low energy transmutation of nuclear waste. This theoretical paper proposes the use of low energy nuclear transmutation (LENT) to reduce or eliminate the growing inventory of radioactive waste, a problem for which we currently have no solution.

E.E.
May 2012

IDEAS & EXPERIMENTS

1. Appearance of Silicon and Metals in Pure Graphite

ABSTRACT

Researchers at Quantum Rabbit LLC (QR) in the USA have repeatedly seen the appearance of silicon and a variety of metals in a pure graphite matrix. In a series of studies conducted since November 2006 at the QR lab in Bellows Falls, Vermont, treated graphite powder shows permanent magnetic activity plus the presence of metals in the parts per million ranges.

METHOD

Non-metallic graphite powders (scientific grade 99.999% pure) are placed in a pure (99.999%) graphite crucible. The powders are charged with 36 volts of direct current through a pure (99.999%) graphite rod. The crucible is connected to the negative pole, the rod to the positive pole of a power pack consisting of three 12-Volt solar-charged batteries. The powders receive between 100 to 200 strikes from the charged rod. Upon cooling the powders are tested for magnetic properties with a neodymium magnet before packaging and shipping to an outside lab for EDS and ICP analysis.

RESULTS

The powders display apparent magnetic activity following the above treatment. Moreover, magnetic activity remains in the powders six months after treatment, suggesting the effect is permanent. Treated graphite shows the presence of magnetic iron at a level of up to 1.6% by weight. A typical sample (ICP analysis by New Hampshire Materials Laboratory, August 9, 2007) shows the appearance of silicon and metals in treated graphite as follows:

Element	Composition Sample (ppm*)
Silicon	10,500
Magnesium	1800
Iron	4700
Copper	4200
Aluminum	7800
Titanium	440
Sulfur	580
Potassium	1000

*Parts per million

Please note the presence of silicon at 1.5% in the treated graphite. No silicon was used in the graphite materials (rod, crucible, powder) employed in the test.

In addition to changes in the composition of the graphite powder used in the tests, changes have been noted in the pure graphite rods used in the experiments. In a study conducted in October 2007, a shiny metallic "bubble" appeared on the striking surface of the rod. Upon analysis, the rod was found to contain the following metals:

Element	Composition Sample (ppm*)
Scandium	35
Iron	640
Cobalt	160
Nickel	1120

*Parts per million

CONCLUSION

Quantum Rabbit research has repeatedly demonstrated that carbon-based materials (pure graphite powder and rods) can develop magnetic properties when exposed to carbon arcing.

Consistent presence of iron and other metals in treated graphite suggest the possibility that the charging process generates low-energy nuclear reactions (transmutations) that result in the appearance of new elements. The sudden charge of electricity may temporarily neutralize the mutual repulsive force existing between two plus charged nuclei. This instantaneous breach may allow the centripetal Casimir force generated by the vacuum/ether to force the nuclei to fuse and form a heavier atom.

The formulas presented below describe possible low-energy fusion reactions that could explain the presence of the new elements in the treated graphite.

Possible Low Energy Nuclear Reactions in Treated Graphite*

Magnesium-24
$_{12}C + _{12}C \rightarrow _{24}Mg$

Calcium-40
$_{24}Mg + _{16}O \rightarrow _{40}Ca$

Aluminum-27
$_{12}C + _{15}N \rightarrow _{27}Al$

Titanium-44
$_{28}Si + _{16}O \rightarrow _{44}Ti$

Silicon-28
$_{12}C + _{16}O \rightarrow _{28}Si$

Iron-56
$2(_{12}C + _{16}O) \rightarrow _{56}Fe$ (- 2 protons)

Potassium-39
$_{26}Mg + _{14}N \rightarrow _{39}K$

Scandium-45
$_{30}Si + _{15}N \rightarrow _{45}Sc$

*Please note that the gases involved in these reactions, oxygen (O) and nitrogen (N), are from the atmosphere.

Our research confirms earlier studies conducted by George Ohsawa and associates and reported in *Infinite Energy*. Further research under rigorously controlled conditions, is needed to determine whether or not low-energy nuclear reactions are taking place in treated graphite materials.

Source: Edward Esko, "Production of Metals in Non-Metallic Graphite," *Infinite Energy* Issue 78, 2008.

Charging graphite powder with graphite rod and crucible.

Charging graphite powder with graphite rod and copper plate.

The carbon arc produces a brilliant flash, similar to a lightning strike.

2. Appearance of Argon in Oxygen-Helium Plasma

ABSTRACT

In gas studies conducted at Quantum Rabbit (QR) lab in Nashua, New Hampshire, USA, in November and December 2005, QR researchers recorded the anomalous appearance of argon in vacuum tubes containing oxygen and helium plasma. Results were monitored and recorded by residual gas analyzer (RGA).

METHOD

Experiment 1: Oxygen plus Helium

4 parts O + 1 part He
Pressure O_2 = 2x He

In the first experiment, the QR vacuum tube was filled with pure oxygen, isolated, and pumped down to approximately 4 torr. Plus/minus stainless electrodes located at opposite ends of the horizontal tube were electrified to the point at which oxygen plasma was created. Pure helium was added as a second fill gas to 4 + 2(1.08) = 6.16 torr so as to maintain the 4 to 1 ratio between oxygen and helium, determined by the respective mass numbers (O-16/He-4) of the gases.

Experiment 2: Helium plus Oxygen

1 part He + 4 parts O
Pressure O_2 = 2x He

In the second experiment, which is the reverse of the above, the QR vacuum tube was filled with pure helium, isolated, and pumped down to approximately 2 torr. The stainless electrodes were electrified so as to spark and maintain helium plasma. Pure oxygen was added as a second fill gas to 2 + 4(1.08) = 6.32 torr, again, to maintain the 1 part He to 4 part O ratio determined by their respective mass numbers.

RESULTS

RGA analysis at the beginning of each test showed the presence of the first fill gases, oxygen in Experiment 1 and helium in Experiment 2. The second fill gases, helium in Experiment 1 and oxygen in Experiment 2, were also noted. In both experiments, values corresponding to the mass assignment of argon (Source: Mass Assignment Table from MKS Instruments, manufacturer of the RGA) appeared on the RGA readout following admission of the second fill gas. No argon was used in either test. These values are shown below:

Mass	Ion	Parent Substance
18	Ar ++	argon-36
20	Ar ++	argon-40
36	Ar +	argon-36
40	Ar +	argon-40

CONCLUSION

Although preliminary, these results suggest the possibility of a low energy nuclear reaction in which one atom of helium fuses with a molecule of oxygen to produce argon:

$$2(_8O_{16/18}) + _2He_4 \rightarrow _{18}Ar_{36/40}$$
$$_2He_4 + 2(_8O_{16/18}) \rightarrow _{18}Ar_{36/40}$$

Placing a target gas under vacuum, charging it with electricity, then admitting a catalyst gas may cause the mutually repulsive force existing between their respective nuclei to become instantaneously neutralized.

This sudden breach may allow the centripetal Casimir force generated by the vacuum/ether to cause the nuclei to fuse and form a heavier atom. In this case two lighter atoms, oxygen and helium fuse to produce the heavier argon atom.

These results are of course preliminary. Further research is needed to establish whether low energy nuclear reactions are in fact produced by the above experiments.

Source: Edward Esko, "Appearance of Argon in Oxygen/Helium Plasma," *Infinite Energy* Issue 81, 2008.

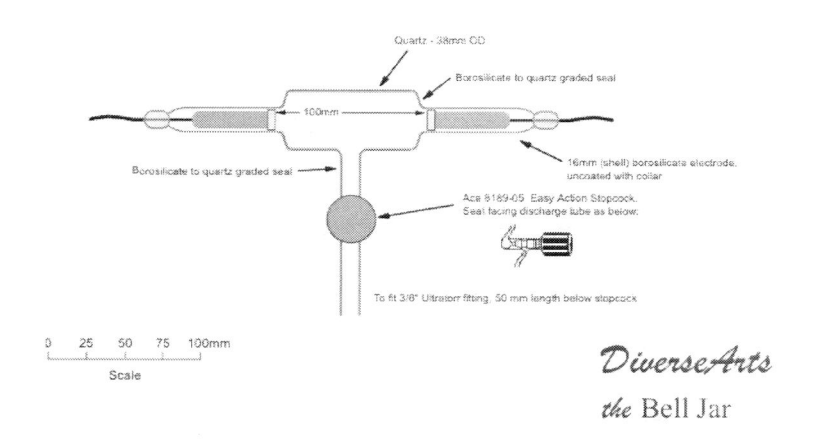

Design of vacuum tube used in Quantum Rabbit (QR) gas experiments.

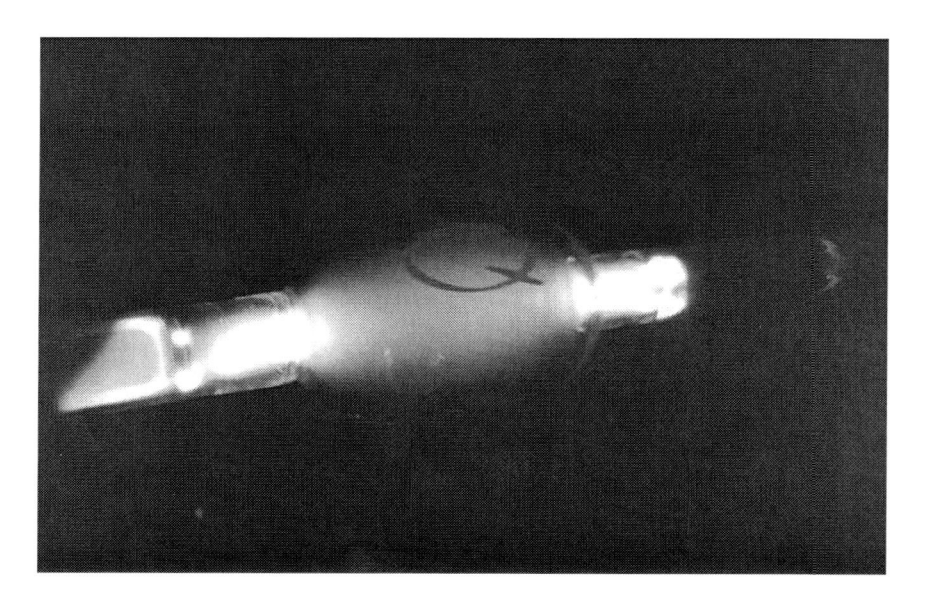

Oxygen plasma in QR vacuum tube.

Helium plasma in QR tube.

3. Appearance of Copper on a Stainless Electrode

ABSTRACT

In a study conducted at Quantum Rabbit (QR) lab in Nashua, New Hampshire, USA, on May 30, 2008, QR researchers noted the anomalous appearance of what seemed to be copper on a stainless steel anode used in a lithium/oxygen vacuum test. The anode was sent to an outside lab, analyzed by ICP (Inductively Coupled Plasma Atomic Emission Spectroscopy), and found to contain trace amounts of copper.

BACKGROUND

In lithium studies conducted on February 29, 2008 and May 2, 2008 at the QR lab in Nashua a small piece of lithium was cut from a lithium rod 12.7 mm in diameter and approximately 165 mm long and placed in a glass vacuum tube. The lithium rods were from Alfa Aesar (Stock 10773/Lot E31S039). The Certificate of Analysis (CofA) of the lithium rods showed the following values:

Composition of Lithium Rod*
Lithium	100% (minus the trace elements below)
Sodium	31
Calcium	17
Iron	1
Silicon	4
Chlorine	16
Nitrogen	102
Potassium	36

*Values given in parts per million (ppm) unless otherwise noted.

The tube was pumped down to vacuum and backfilled with pure oxygen to about 3 torr. Stainless steel electrodes at either end of the tube were electrified by batteries charged by a HUBERT® portable solar generator. The samples were heated from the outside by a hand torch, causing them to vaporize.

Lithium samples were retrieved, stored in mineral oil and sent to New Hampshire Materials Laboratory (NHML) for EDS and ICP analysis. The February 29th lithium test sample (NHML File 24901) was found to contain sodium at 0.94% and potassium at 0.14%, a substantial increase over the values in the CofA. ICP analysis of the May 2 lithium test samples showed similar results (NHML File 25125):*

Element	Control	Test 1	Test 2
Sodium	50	213	8900
Potassium	5	5	6

*Parts per million

Upon review, QR researchers determined that the increase in sodium was due to one or both of two factors:

1. A low energy nuclear reaction between lithium and oxygen producing the heavier element sodium:

$_7Li + {}_{16}O \rightarrow {}_{23}Na$
Lithium-7 + oxygen-16 → sodium-23

2. The release of sodium contained in the Pyrex (borosilicate) glass used in the tube. Wayne Martin, of M & M Glassblowing in Nashua, fabricated the glass tubes used in these studies. Wayne pointed out that the tubes were composed primarily of SiO_2 and B_2O_3. The glass also contained anywhere from 2% to 3% Al_2O_3, and most significant to our results, sodium and potassium oxides (Na_2O+K_2O) at up to 4.2%. Recent studies have shown that when heated, as the QR tubes were, borosilicate glass would release sodium.

Another series of lithium tests was set for May 30, with the goal being to control for sodium. A new tube was designed so that the 60 mm center section would be made of quartz. The type of quartz used in the new tube contained a very small percentage of sodium (0.7 ppm), much less than that found in borosilicate. We felt comfortable that this trace amount wouldn't have a noticeable effect on the outcome of the test.

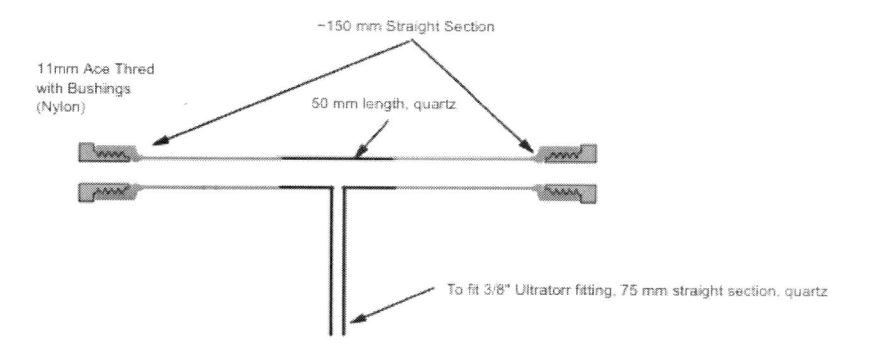

Tube design used in QR metal vapor studies. Stainless electrodes were placed at either end and test material placed in the quartz midsection between anode and cathode.

METHOD

A small piece of lithium was cut from a fresh lithium rod and placed in the middle of the quartz section of the QR tube (see illustration.) Stainless electrodes (anode and cathode) were inserted at each end of the tube so that the tip of the anode came into contact with the lithium pellet. The tube was pumped down to vacuum and backfilled with pure oxygen to approximately 3.5 torr. A charge was passed through the electrodes producing a glow discharge. The sample was heated from the outside by a hand torch, at which point the metal appeared to vaporize. After several minutes, the electricity was shut off, the tube returned to atmosphere, and the sample allowed to cool.

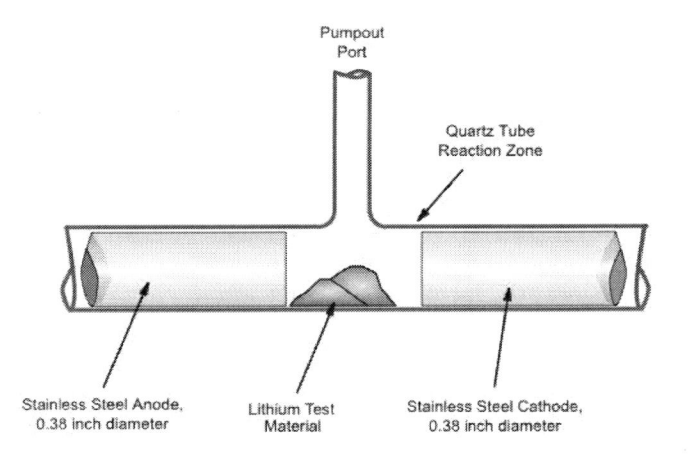

Close-up of the stainless electrodes and lithium sample used on May 30.

RESULTS

The anode, cathode, and lithium pellet were removed from the tube and visually inspected. At the tip of the anode in the region of the anode that had been in contact with the lithium pellet, there appeared a copper-colored residue. This was noted and the samples packaged and sent to New Hampshire Materials Laboratory for EDS and ICP analysis.

ICP analysis showed no detectable presence of sodium. This result added credence to the theory that the sodium detected in the earlier lithium tests had come from the borosilicate glass rather than from a low energy nuclear reaction between lithium and oxygen.

However, consistent with the visual examination, the anode sample was found to contain 1500 ppm copper (NHML File 25237). As revealed in the CofA, the test sample of lithium contained no copper. The stainless anode also did not contain copper. The electrodes were composed primarily of iron. Their composition is as follows (shown in percentages):

The dark irregular pellet is the lithium metal used in the May 30 test.

Carbon	0.15 max
Chromium	17 - 19
Iron	Balance
Manganese	2 max
Nickel	8 - 10
Phosphorus	0.045 max
Silicon	1 max
Sulfur	0.03 max

Source: Diverse Arts/The Bell Jar

The quartz used in the tube contained a very small amount of copper—0.05 ppm—in addition to other trace elements in the ppm range. However, the amount of copper found on the electrode—1500 ppm—was *thirty thousand* times that amount. The experiment apparently produced an unexpected, dramatic, and unexplained increase in the element copper.

CONCLUSION

This result suggests the possibility of a low energy nuclear reaction between the pure lithium pellet and the iron in the stainless anode. During the experiment, atoms of lithium may have reacted with atoms of iron to produce atoms of copper:

$_7Li + {}_{56}Fe \rightarrow {}_{63}Cu$
Lithium-7 + iron-56 → copper-63

Placing the lithium in contact with the iron-rich anode, pumping down to vacuum, admitting a catalyst gas (in this case oxygen), and electrifying the tube may have caused the mutually repulsive force existing between the lithium and iron nuclei to become instantaneously neutralized. This sudden breach may have allowed the centripetal Casimir force generated by the vacuum/ether to cause the nuclei to react and form a heavier atom. In this case, two lighter atoms—lithium and iron—may have reacted to produce a heavier atom of copper.

Although preliminary, this result justifies further research to determine whether or not a low energy nuclear reaction is indeed taking place in this experiment.

AFTERWORD — VERIFICATION

The QR research team repeated the experiment on Dec. 30, 2008. The results confirmed those of May 30. Copper appeared on the stainless anode at 262 ppm and on the stainless cathode at 315 ppm. New Hampshire Materials Lab, File 25985, conducted analysis of the stainless electrodes.

Source: Edward Esko, "Appearance of Copper on a Stainless Electrode," *Infinite Energy* Issue 86, July/August 2009.

4. Appearance of Palladium on a Zinc Anode

ABSTRACT

In a study conducted at Quantum Rabbit (QR) lab in Nashua, New Hampshire, USA, on September 30, 2008, the QR research team, myself, together with Alex Jack and Woody Johnson conducted a vacuum discharge test with a zinc anode, copper cathode, and pure sulfur. Oxygen was added as a fill gas. Upon analysis by ICP (Inductively Coupled Plasma Atomic Emission Spectroscopy) at New Hampshire Materials Laboratory, test samples were found to contain significant traces of palladium.

BACKGROUND

Earlier studies at QR labs have suggested the possibility of inducing low energy nuclear reactions under relatively low temperatures and pressures, and with relatively small inputs of energy. In a previous study, QR researchers noted what appeared to be copper residue on the stainless electrode used in a vacuum test.

Upon ICP analysis, the stainless anode was found to contain copper at 1500 ppm. No copper was used in the test. The result suggested the possibility of a low energy nuclear reaction, in which iron in the stainless electrode reacted with the lithium pellet used in the test.

With the copper experiment as background, I formulated a new experiment designed to test the hypothesis that low energy nuclear reactions are taking place in QR vacuum discharge tests. I predicted that, under the right conditions, similar to those in the previous appearance of copper test, an atom of zinc could be made to react with an atom of sulfur to form palladium. An experiment designed to test this hypothesis was scheduled for September 30, 2008 at the QR lab in Nashua.

In a September, 16, 2008 e-mail to K.J. Thomas at the renowned Cavendish lab in England, I expressed my hope for the experiment as well as my desire to participate in cooperative research on low energy transmutation:

"We would very much like to discuss the possibility of conducting a series of studies at your labs at Cavendish. We feel your facilities would allow for more precise before and after analysis, more careful control of test samples, and could extend the research to the next level, that being to detect whether the isotope distribution in test materials matches that found in nature or is unique and whether or not the low energy transmutation of elements produces energy in the form of charged particles.

"Of the various lines we have pursued, the possibility that electrode composition can be changed under vacuum and with particular catalysts is perhaps the most intriguing "We have scheduled additional tests in this series at our New Hampshire lab. We plan to test zinc electrodes under vacuum in combination with sulfur and with oxygen as the fill gas. Our goal is to see whether the zinc electrodes and sulfur and oxygen catalysts can be prompted to form palladium."

METHOD

In the Sept. 30 test, we utilized the tube design employed in the lithium-iron-copper test described above. The borosilicate glass tube was 150 mm long, with a 50 mm quartz midsection. A 3/8-inch diameter quartz straight section, perpendicular to the tube and 75mm in length, connected the midsection with the vacuum manifold, and served as the entrance for a pure oxygen backfill.

We had ordered a high purity copper rod from Alfa Aesar and cut the rod to create the anode and cathode. This Puratronic® rod is composed of 99.999% pure copper (metals basis.) We also ordered a high purity zinc rod (99.9999% metals basis), and used the rod to create a pure zinc insert pressed into a 17/64-inch diameter cavity in the center of the copper anode.

First, we inserted the copper cathode in one end of the tube. Pure sulfur was placed in the quartz midsection. The sulfur, listed in the Alfa Aesar catalogue as "sulfur pieces," was actually a coarse yellow powder that was difficult to maneuver into the tube. However, after persistent attempts, our vacuum consultant succeeded in placing a sufficient quantity in the tube.

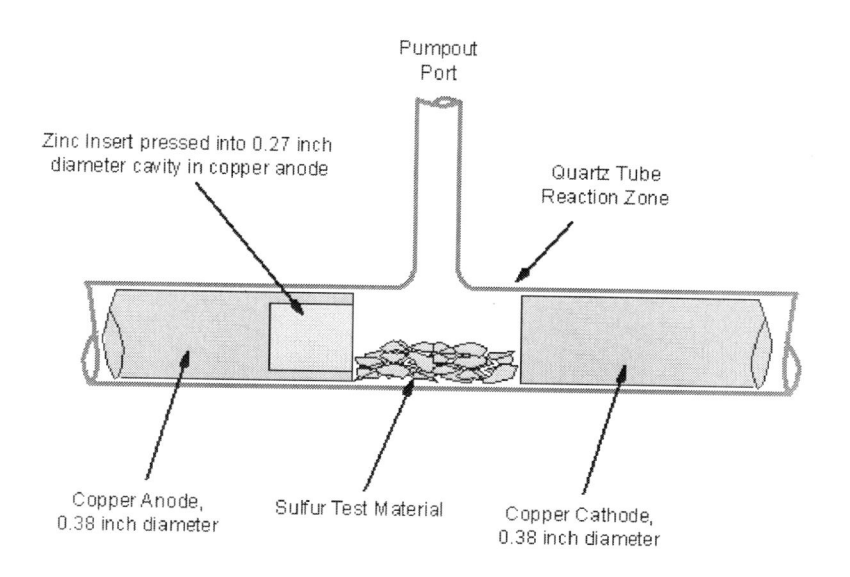

Quantum Rabbit tube and electrode configuration used in the Sept. 30 test.

Like the copper electrodes, the sulfur was labeled Puratronic® and certified by Alfa Aesar as being 99.999% pure (metals basis.)

After placing the sulfur in the tube, the zinc-tipped anode was inserted at the opposite end, and the electrodes connected to the power supply. Both anode and cathode were in contact with the sulfur powder. The tube was pumped down to vacuum and backfilled with oxygen to approximately 3.5 torr. Electricity was passed through the electrodes, producing an electric arc and glow discharge. The glow had a distinctive blue-white hue. A portion of the sulfur had apparently vaporized. A digital camera recorded the sulfur glow. After several minutes, the electricity was turned off, the vacuum pumps disconnected, and the sample allowed to cool.

PALLADIUM RESULTS

Three samples were retrieved from the tube: the zinc-tipped anode, the copper cathode, and a blackened residue from the center of the tube. The anode and cathode tips had undergone a noticeable change during the experiment, with the presence of an undetermined residue on their surface. All three samples were packaged and sent to New Hampshire Materials Laboratory for ICP analysis.

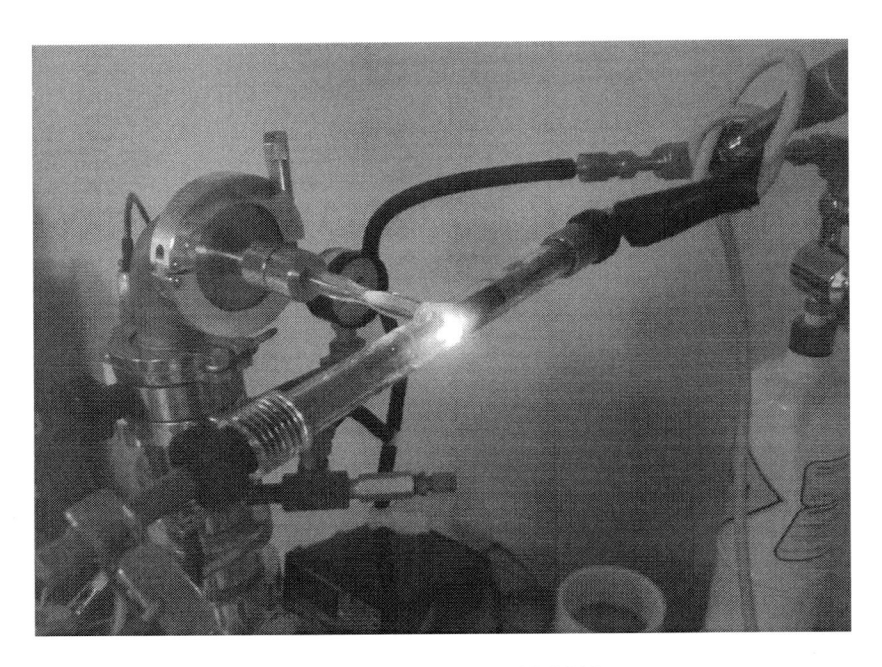

Sulfur discharge glow in QR tube Sept. 30 2008.

ICP results came back a month later (NHML File 25670a dated October 30, 2008). The results appeared to confirm my prediction about the appearance of palladium following a possible low energy nuclear reaction between zinc and sulfur. Here are the differences in the amount of palladium detected in the test samples before and after the experiment. The palladium (Pd) values shown in the "Before" column are from the Certificate of Analysis (CofA) provided by Alfa Aesar for each sample. They reflect the composition of the samples prior to the test. The Pd values in the "After" column are from the NHML ICP analysis report cited above and reflect composition values following the test.

	Pd Before	Pd After
Zinc anode	ND	50 ppm*
Copper cathode	<0.002 ppm	40 ppm
Sulfur residue	ND	91 ppm

*Parts per million

According to the Certificates of Analysis, there was no trace (ND) of palladium in the zinc anode (Stock 12718/Lot LI4R006) and sulfur powder (Stock 10755/Lot D03S015). The amount detected in the copper cathode was cited at 0.002 ppm (Stock 10156/Lot L08R011). After the test, the amount of palladium on the copper cathode jumped to 40 ppm. Moreover, the experiment apparently led to a sudden and dramatic appearance of palladium on the zinc anode tip and in the residue; from zero to 50 ppm on the anode, and zero to 91 ppm in the residue.

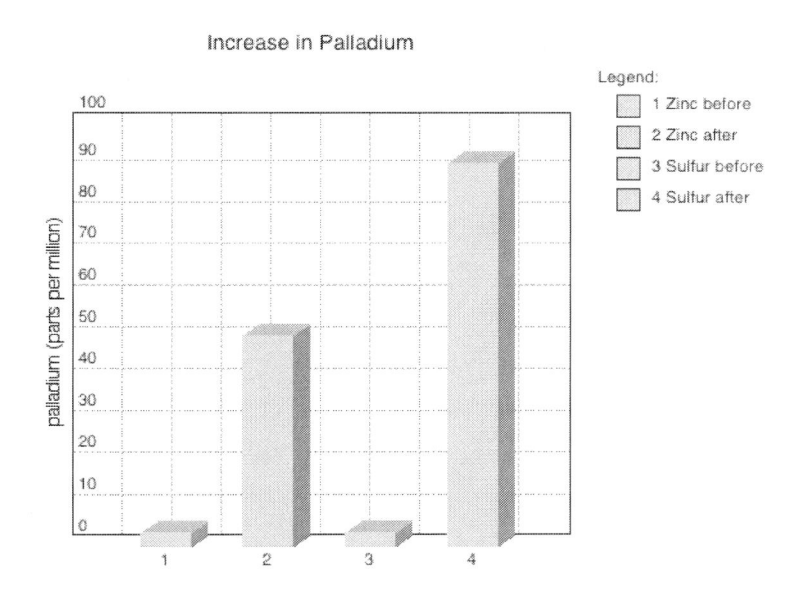

Increase in Palladium

Legend:
- 1 Zinc before
- 2 Zinc after
- 3 Sulfur before
- 4 Sulfur after

Level of palladium on the zinc anode before (1) and after (2) the Sept. 30 test, and in the sulfur residue before (3) and (4) after the test.

CONCLUSION

These results may confirm the predictive power of the quantum conversion theory of low energy transmutation. As predicted, during the experiment, atoms of zinc may have reacted with sulfur to form palladium (elements shown with atomic numbers):

$_{30}Zn + {}_{16}S$ (with oxygen catalyst) $\rightarrow {}_{46}Pd$

Placing sulfur in contact with the zinc anode, pumping down to vacuum, admitting oxygen as a catalyst, and electrifying the tube may have caused the electrical repulsive force existing between two plus charged nuclei—zinc and sulfur—to become temporarily neutralized. Such temporary neutralization may have allowed the omnipresent centripetal Casimir force to cause zinc and sulfur to react and form of a heavier atom of palladium. As we see in the chart below, these three elements, zinc, sulfur, and palladium, have numerous isotopes. It would be most productive to investigate whether the isotope distribution of the palladium detected in the test matches that of palladium found in nature. Hopefully, this can become the focus of future research.

Isotope Matchups

$_{32}S + {}_{70}Zn \rightarrow {}_{102}Pd$
$_{34}S + {}_{68}Zn \rightarrow {}_{102}Pd$
$_{34}S + {}_{70}Zn \rightarrow {}_{104}Pd$
$_{36}S + {}_{66}Zn \rightarrow {}_{102}Pd$
$_{36}S + {}_{68}Zn \rightarrow {}_{104}Pd$
$_{36}S + {}_{70}Zn \rightarrow {}_{106}Pd$

There were other anomalies detected in the test samples. Here is the full ICP analysis from NHML.

Zinc Anode

Used in Test
Copper	1.27%
Sulfur	1.75%
Zinc	1 ppm

Detected in Treated Sample
Strontium	0.5 ppm
Palladium	50 ppm
Chromium	8 ppm

Copper Cathode

Used in Test
Copper	64.2%
Sulfur	117 ppm
Zinc	2 ppm

Detected in Treated Sample
Strontium	2 ppm
Palladium	40 ppm
Chromium	2 ppm

Residue

Used in Test
Copper	35.7%
Sulfur	7%
Zinc	3 ppm

Detected in Treated Sample
Strontium	1 ppm
Palladium	91 ppm
Chromium	<1 ppm

In addition to palladium, note the consistent presence of strontium and chromium in the test samples. According to the CofA, strontium was not present in the zinc anode or sulfur powder prior to the test. Only a minute amount—0.0001 ppm—is listed in the CofA for the copper rod. That is a vastly smaller quantity than the 2 ppm found on the copper cathode after the test. In the case of chromium, none is listed in the CofA for the zinc rod. The CofA for the sulfur powder shows chromium at less that 0.5 ppm. The CofA for the copper rod lists chromium at 0.002 ppm. Both quantities are far less than the 8 ppm found on the zinc anode after the test.

The anomalous appearance of these trace metals in the test samples may be another example of the predictive power of the quantum conversion theory. Prior to the experiment, I predicted that, through the process of low energy transmutation, zinc could be converted to strontium:

$$_{70}Zn + {}_{16}O \rightarrow {}_{86}Sr$$
Zinc-70 + oxygen 16 → strontium-86

I also predicted that an atom of sulfur could be induced to react with an atom of oxygen to form an atom of chromium:

$$_{34}S + {}_{16}O \rightarrow {}_{50}Cr$$
Sulfur-34 + oxygen-16 (fill gas) → chromium-50

Having worked out these formulas in advance, I asked Tim Kenney, director of laboratory services at NHML, to scan for strontium and chromium, in addition to palladium, in the ICP analysis. Their presence in the test samples was therefore more confirmation than surprise.

In another test done with a zinc anode, copper cathode, sulfur test material, and oxygen backfill conducted at QR lab on December 30, 2008, strontium appeared at 14 ppm on the zinc anode, 4 ppm on the copper cathode, and 1 ppm in the sulfur residue. Chromium appeared at 130 ppm on the zinc anode and 198 ppm in the sulfur residue (NHML File 25985.)

When I addressed the special workshop on transmutation organized by George Miley following ICCF-14 (International Conference on Cold Fusion) in August 2008 in Washington DC, I challenged the audience to conduct experiments similar to the Quantum Rabbit experiments in their laboratories. I wish to repeat that challenge. Although preliminary, the results reported here, and in my previous papers published in *Infinite Energy*, are highly suggestive. My hope is that researchers around the globe will initiate independent investigations with the potential to replicate and add credence to these original formulas and novel results. Feel free to contact me if you would like our theoretical and technical support for your research.

Source: Edward Esko, "Appearance of Palladium on a Zinc Anode," *Infinite Energy* Issue 87, September/October 2009.

5. Appearance of Tin on a Silver Anode

ABSTRACT

ICP (Inductively Coupled Plasma Atomic Emission Spectroscopy) analysis of a pure silver anode used in a vacuum study conducted at Quantum Rabbit (QR) lab in Nashua, New Hampshire, USA, on September 30, 2008, revealed the anomalous appearance of tin.

BACKGROUND

Following a vacuum discharge study conducted at the QR lab on May 30, 2008, in which copper was found to have appeared on the tip of a stainless anode, I formulated an experiment designed to test the hypothesis that an atom of silver could be induced to react with an atom of lithium to form an atom of tin. The experiment was conducted on the same day (September 30) as the experiment in which palladium was found to have appeared on a zinc anode.

The silver-tin experiment was conducted by QR researchers Alex Jack, Woody Johnson, and me, with the assistance of our vacuum consultant, Steve Hansen. Our master glassblower, Wayne Martin, of M & M Glassblowing in Nashua, had fabricated the vacuum tube according to my design.

Prior to the experiment, I had sent an e-mail to K.J. Thomas at the Cavendish Lab in the UK. In that September 16 e-mail, I described the experiment and stated our primary goal: "In the test, we plan to see whether silver and lithium can be prompted to produce tin."

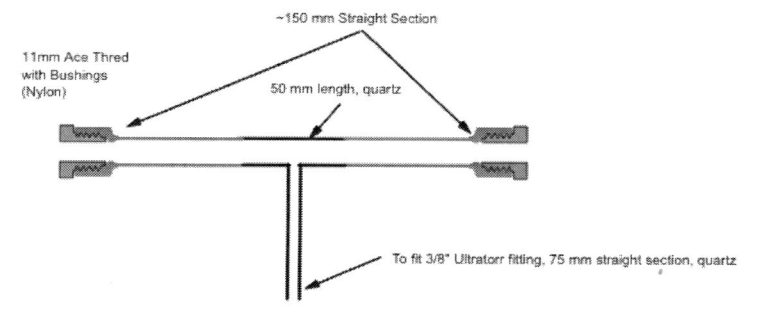

Quantum Rabbit vacuum tube used in the Sept. 30 test.

METHOD

For this test, we employed the tube design used in the metal vacuum tests described above. The borosilicate glass tube was 150 mm long, with a 50 mm quartz midsection. A 3/8-inch diameter quartz strait section, perpendicular to the tube and 75mm in length, sealed the connection to the vacuum manifold, and served as the entryway for the pure oxygen employed as a fill gas. The anode and cathode were made of a pure copper rod from Alfa Aesar. The rod is listed in the catalogue as Puratronic® (Stock 10156/Lot L08R011) and is composed of 99.999% pure copper (metals basis). There are 62 trace elements listed in the CofA (certificate of analysis) for the copper rod, most in fractions of a part per million. The list is too long to present here. Readers who are interested in obtaining the CofA for the copper rod can contact Alfa Aesar directly.

A pure silver rod was also ordered from Alfa Aesar (Stock 11473/Lot E14M22). The high purity silver rod is listed as Premion and certified at 99.999% pure. The analysis of the rod provided by the CofA shows the following trace elements, in addition to silver (Ag):

Composition of Silver Rod*

Ag	Major
Pt	1
B	ND
Ni	ND
Zn	1
Cu	2
Pd	ND

Fe	4
Pb	ND
Au	ND
Al	ND
Mg	ND
Sn	ND

*Values are given in parts per million (ppm) unless otherwise noted. ND: Sought but not detected

Please note the value given for tin (Sn): sought but not detected. This value serves as the primary point of comparison with the values detected at the outcome of the experiment.

The silver rod was used to create a silver inset—0.27 inches in diameter—at the center of the copper anode. First, the copper cathode was inserted at one end of the tube. A small piece of lithium was cut from a pure lithium rod and placed in the quartz midsection. The lithium rod was from Alfa Aesar (Stock 10773/Lot E31S039). The Certificate of Analysis (CofA) of the lithium rod showed the following values:

Composition of Lithium Rod*

Lithium	100%**
Sodium	31
Calcium	17
Iron	1
Silicon	4
Chlorine	16
Nitrogen	102
Potassium	36

*Values given in parts per million (ppm) unless otherwise noted.
**Minus the elements shown in the analysis

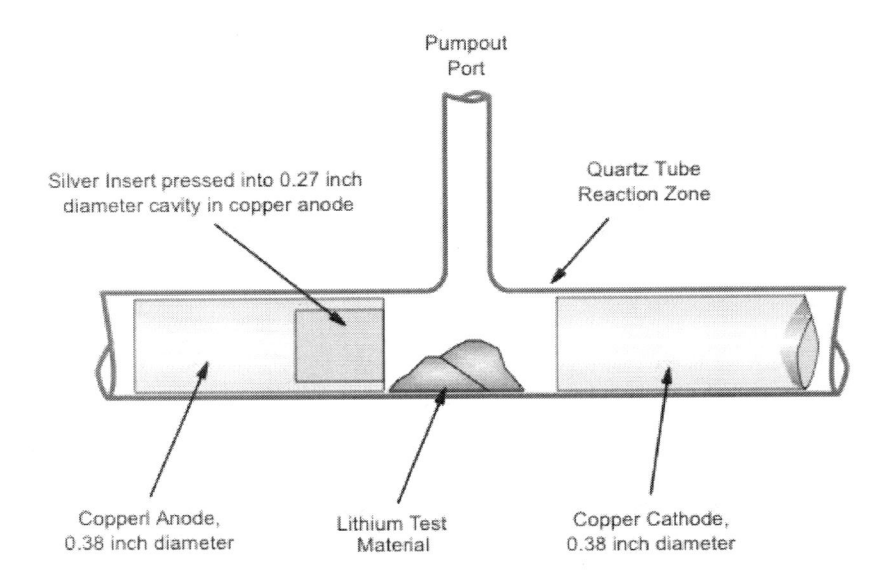

Tube and electrode configuration used in the Sept. 30 test.

Following insertion of the lithium, the silver-tipped anode was inserted in the tube so that the silver anode came into contact with the lithium. The electrodes were connected to the power supply, and the tube pumped down to vacuum and backfilled with pure oxygen to approximately 3.5 torr. A charge was passed through the electrodes producing a glow discharge. The sample was heated from the outside by a hand torch, at which point the metal appeared to vaporize, giving off the ruby red color characteristic of lithium. After several minutes, the electricity was shut off, the tube returned to atmosphere, and the sample allowed to cool.

RESULTS

The anode, cathode, and lithium pellet were removed from the tube and visually inspected. The tips of both electrodes had undergone obvious changes, as had the lithium pellet. These changes were noted and the samples packaged and sent to New Hampshire Materials Laboratory (NHML) for EDS and ICP analysis.

Woody Johnson (left) and Steve Hansen observe lithium vapor in the QR vacuum tube, Sept. 30, 2008.

The Test Report (NHML File 25670a) revealed the following values, as the result of chemical analysis by ICP:

<u>Analysis Results</u>

Silver Anode
Copper	65 ppm
Tin	2 ppm
Germanium	<144 ppm
Silver	38.12%

Copper Cathode
Copper	89.19%
Germanium	<38 ppm

Lithium Pellet
Copper	1050 ppm
Tin	3 ppm
Germanium	<70 ppm

CONCLUSION

As predicted by Quantum Conversion Theory, tin was found in the silver anode and lithium pellet. Although the amounts of tin detected in the treated samples are small, in the parts per million ranges, they represent a significant increase over the amounts present prior to the experiment. If we refer to the analyses of the silver and lithium rods used in the test, we see that no traces of tin are present. Moreover, there are no traces of tin in the borosilicate and quartz glass used in the vacuum tube, according to analysis provided by Wayne Martin. The only possible source of tin in the experiment was the copper rod. However, the amount of tin in the copper rod (listed in the CofA) is far less than that found in the materials after the test: less than 0.01ppm in the copper rod before the test, versus 2ppm in the silver anode and 3ppm in the lithium pellet after the test, a 200- to 300-fold increase. The dramatic increase in tin suggests the possibility of a low energy nuclear reaction between lithium and silver, with oxygen as the catalyst (elements shown with atomic numbers):

$$_{47}Ag + {}_3Li \text{ (with oxygen catalyst)} \rightarrow {}_{50}Sn$$

Placing lithium in contact with the silver anode, pumping down to vacuum, admitting a catalyst gas (oxygen), and electrifying the tube may have caused the repulsive force existing between positively charged silver and lithium nuclei to become temporarily neutralized. This temporary breach may have allowed the centripetal Casimir force generated by the vacuum/ether to cause the nuclei to react and form a heavier atom. In this case, two lighter atoms—lithium and silver—may have reacted to form a heavier atom of tin.

As we see below, these three elements—silver, lithium, and tin—have isotopes that could provide pathways for low energy transmutations.

Isotope Pathways

Silver	Lithium	Tin
$_{109}Ag$	$_6Li$	$_{115}Sn$
$_{107}Ag$	$_7Li$	$_{114}Sn$
$_{109}Ag$	$_7Li$	$_{116}Sn$

Increase in Tin

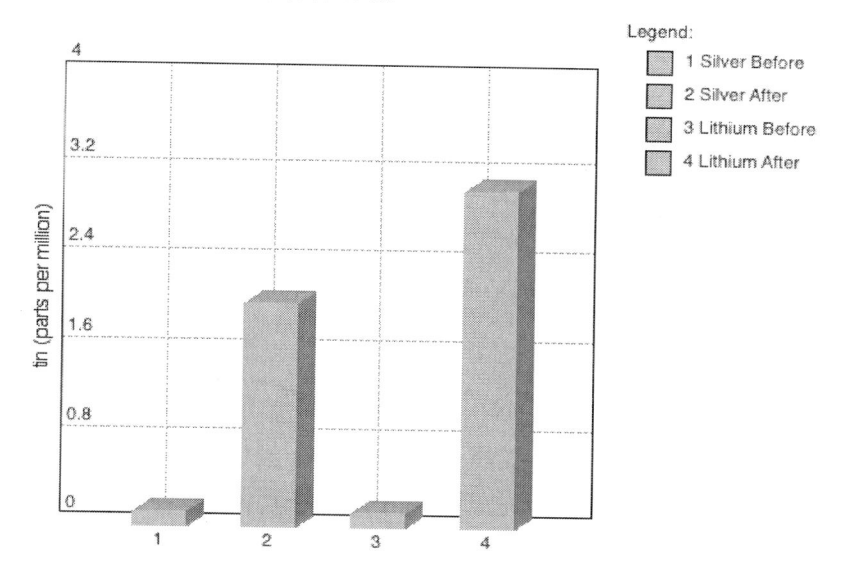

(Bar chart showing tin in parts per million for four categories)
- 1 Silver Before: ~0
- 2 Silver After: ~1.8
- 3 Lithium Before: ~0.1
- 4 Lithium After: ~2.9

Legend:
- 1 Silver Before
- 2 Silver After
- 3 Lithium Before
- 4 Lithium After

The chart shows the increases in tin in parts per million on the silver rod before (1) and after (2) the Sept. 30 test; and in the lithium metal before (3) and after (4) the test.

Although preliminary, these results justify further research to determine whether or not a low energy nuclear reaction between silver and lithium is indeed taking place in the experiment.

THE GERMANIUM QUESTION

In addition to the above result, I had predicted the possibility that germanium (Ge) would be produced during the experiment. I expressed this possibility in my September 16 e-mail to K.J. Thomas:

"We plan to use a copper anode and silver cathode under vacuum, in combination with lithium and oxygen as the fill gas...to see whether copper and lithium can be prompted to produce trace amounts of germanium."

The prediction was based on the Quantum Conversion formula:

$$_{63}Cu + {_7}Li \text{ (with oxygen catalyst)} \rightarrow {_{70}}Ge$$
Copper-63 + lithium-7 → germanium-70

It was for this reason that I asked Tim Kenny at New Hampshire Materials Lab to test for germanium in the samples. However, as we see in the analysis report, the germanium result is inconclusive. The amounts indicated in the report represent the detection limits for germanium, which, in comparison to many other elements, are high. Therefore, we weren't able to say with precision whether or not germanium was present in the test samples.

However, in vacuum tests run on December 30, 2008, in which copper electrodes, lithium test material, and oxygen backfill were employed, as in the Sept. 30 test, germanium appeared as follows (ICP analysis by New Hampshire Materials Lab, File 25985):

Copper Anode	Copper Cathode	Lithium Residue
260 ppm	50 ppm	106 ppm/154 ppm (in a second test)

These results suggest the possibility of the low energy nuclear reaction postulated above.

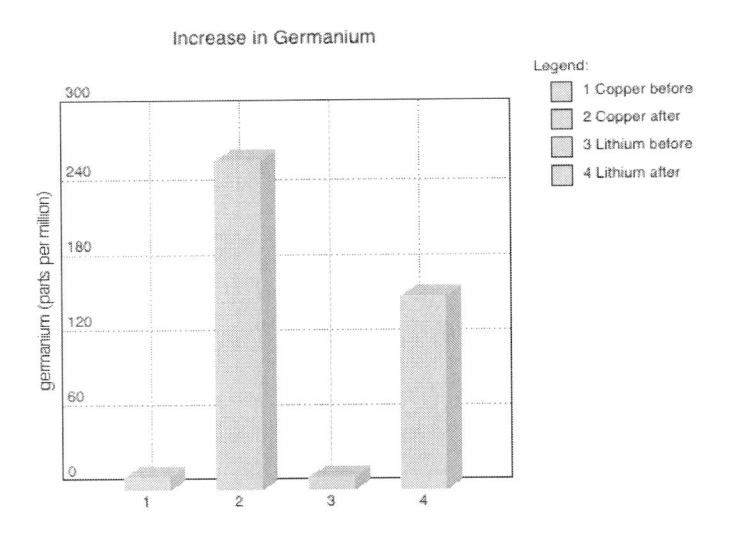

Increase in germanium in the Dec. 30, 2008 study.

The chart shows the increases in germanium in parts per million on the copper anode before (1) and after (2) the Dec. 30 test; and in the lithium metal before (3) and after (4) the test. The amount in the copper prior to the test was estimated to be 0.02 ppm and 0 ppm in the lithium.

AFTERWORD — VERIFICATION

Perhaps the most dramatic germanium result occurred in a July 30, 2009 test sponsored by the New Energy Foundation and conducted at the Quantum Rabbit lab in Owls Head, Maine. Copper electrodes were used in the test, as was a lithium pellet and pure oxygen backfill. ICP conducted by NHML revealed germanium at 2190 ppm on the copper cathode and 388 ppm on the lithium pellet (NHML File 26657/August 1, 2009), adding further confirmation of a possible low energy nuclear transmutation taking place in Quantum Rabbit studies.

Source: Edward Esko, "Appearance of Tin on a Silver Anode," *Infinite Energy* Issue 88, November/December 2009.

6. Carbon Arc Under Vacuum

ABSTRACT/BACKGROUND
Researchers at Quantum Rabbit (QR) LLC in the United States have repeatedly seen the anomalous appearance of silicon and a variety of metals, including magnesium, aluminum, scandium, iron, cobalt, and nickel in pure graphite powder. These carbon arc studies were conducted in the open air and seem to have produced a cascade of low energy nuclear reactions in which carbon (in the graphite powder and graphite rods used in the arcing process) reacted with oxygen and nitrogen. A summary of this research—"Production of Metals from Non-Metallic Graphite"—was published *in Infinite Energy*, Issue 78, 2008.

On December 30, 2008, QR researchers achieved similar results, notably the appearance of magnesium, silicon, iron and other elements on the surface of the pure graphite rods used in a carbon arc test performed under vacuum. The study took place at the QR lab in Nashua, New Hampshire, USA.

METHOD
Unlike previous carbon arc studies conducted in open air, the Dec. 30 test took place in a specially designed vacuum tube. The tube design was that utilized in previous QR vacuum studies. The borosilicate tube was 150 mm long, with a 60 mm quartz midsection. A 3/8-inch diameter quartz straight section, perpendicular to the tube and 75 mm in length, connected the quartz reaction zone with the vacuum manifold, and served as the entrance for a pure oxygen backfill. None of the elements later found on the electrodes were in the quartz material.

We had ordered high purity graphite rods from Alfa Aesar and used the rods to create the anode and cathode. The graphite electrodes were inserted into aluminum electrodes, 0.38-inch in diameter, designed to be flush with the inner wall of the tube so as to seal the tube to maintain vacuum. The aluminum electrodes were positioned outside the quartz reaction zone so as not to interfere with reactions taking place in the center of the tube.

The graphite rods were ultra high purity, 99.9995% pure (metals basis). Their CofA is as follows (Stock 14754/Lot F27S027):

> Graphite electrode, counter-flat top, 3.05mm dia,, 38.10 mm long, 99.9995% (metals basis)
>
> | Al | ND | Fe | ND | Ni | ND | W | ND |
> |----|----|----|----|----|----|----|----|
> | B | ND | Pb | ND | Si | ND | V | ND |
> | Ca | ND | Mg | ND | Ag | ND | Zn | ND |
> | Cr | ND | Mn | ND | Sn | ND | Zr | ND |
> | Ca | ND | Mo | ND | Ti | ND | | |
>
> Values given in ppm unless otherwise noted
> ND: Not detected

First, the graphite cathode was secured in the tube, and a catalyst of pure sulfur was placed in the quartz midsection. The sulfur, listed in the Alfa Aesar catalogue as "sulfur pieces," was actually a coarse powder. It was labeled as Puratronic® (Stock 10755/Lot D03S015) and certified as being 99.999% pure (metals basis.)

After the sulfur was placed in the tube, the graphite anode was inserted in the opposite end, and the electrodes connected to the power supply. The tube was pumped down to vacuum and back-filled with pure oxygen to approximately 3.5 torr.

Electricity was passed through the electrodes, producing an electric arc and glow discharge with the blue/white characteristic of sulfur. After several minutes, the electricity was turned off, the vacuum pumps disconnected, and the sample allowed to cool.

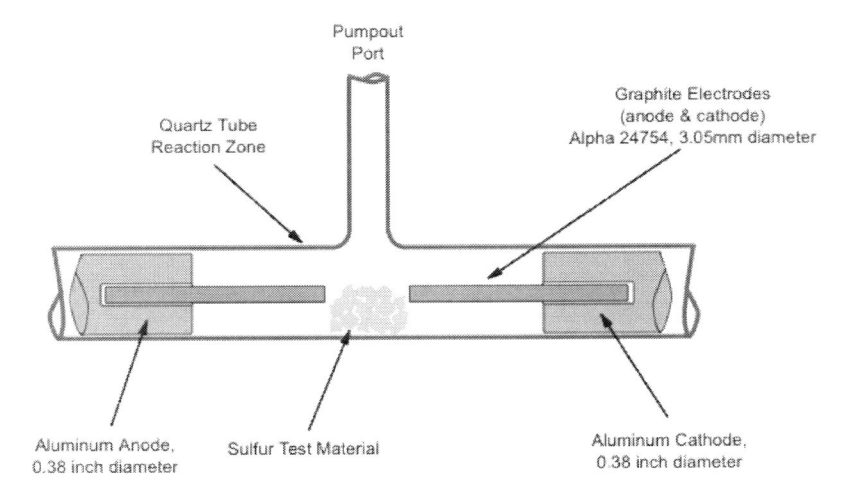

Tube and electrode configuration used in the Dec. 30 test.

The graphite anode and cathode tips had undergone noticeable changes during the arcing process, including the presence of an undetermined residue on their surface. Both graphite electrodes were packaged and sent to New Hampshire Materials Laboratory for analysis by ICP (Inductively Coupled Plasma Atomic Emission Spectroscopy).

ICP results came back on January 9, 2009 (NHML File #25985). Results were consistent with previous carbon arc studies and are presented below.

MAGNESIUM

No magnesium was listed in the CofA for the graphite rods. The CoA for the sulfur pieces listed magnesium at 0.2 ppm. The treated anode came back with a magnesium content of 7 ppm; and the cathode, 8 ppm (see table "Increase in Magnesium.") The presence of magnesium was consistent with open-air studies, one of which (August 9, 2007, NHML File 24067) revealed magnesium at up to 1800 ppm. The appearance of magnesium across a wide spectrum of carbon arc studies suggests a basic nuclear reaction:

$$_{12}C + _{12}C \rightarrow _{24}Mg$$
Carbon-12 + carbon-12 → magnesium-24

Simply put, when exposed to arcing, both in open air and under vacuum, two atoms of carbon may react with each other to form an atom of magnesium.

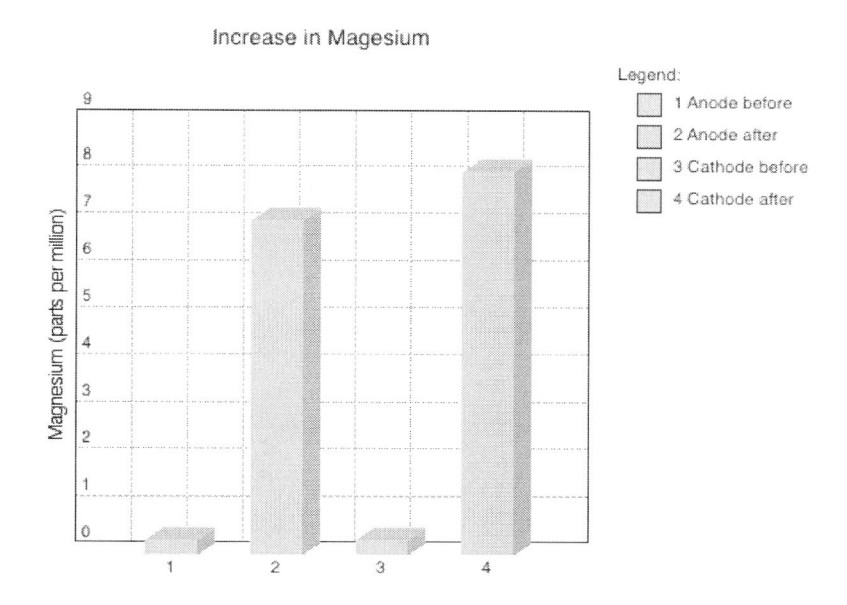

Increase in Magesium

[Bar chart showing Magnesium (parts per million) for: 1 Anode before (~0), 2 Anode after (~7), 3 Cathode before (~0), 4 Cathode after (~8). Legend: 1 Anode before, 2 Anode after, 3 Cathode before, 4 Cathode after]

Levels of magnesium in anode and cathode before and after the test.

SILICON

Together with magnesium, silicon has appeared consistently in carbon arc studies. In the open-air study cited above (NHML File 24067), silicon appeared in treated graphite at 10,500 ppm, or 1.5%. ICP of the treated anode from Dec. 30 revealed silicon at 61 ppm and of the treated cathode at 138 ppm (see table "Increase in Silicon.") No silicon was listed in the CofA for the graphite rods. Analysis of the sulfur pieces revealed silicon at less than 1 ppm.

The consistent appearance of silicon suggests a fundamental relationship existing between carbon and oxygen. As we know from chemistry, carbon and oxygen readily combine to form compounds like carbon dioxide and the organic compounds that form the basis of life.

Apparently, given the proper conditions, the mutual attraction existing between these two elements can prompt them to react and form the heavier atom silicon. The formula for this transmutation is as follows:

$$_{12}C + {}_{16}O \rightarrow {}_{28}Si$$
Carbon-12 + oxygen-16 → silicon-28

Apparently, the transmutation of carbon and oxygen into silicon can be achieved under relatively low pressures, temperatures, and inputs of energy.

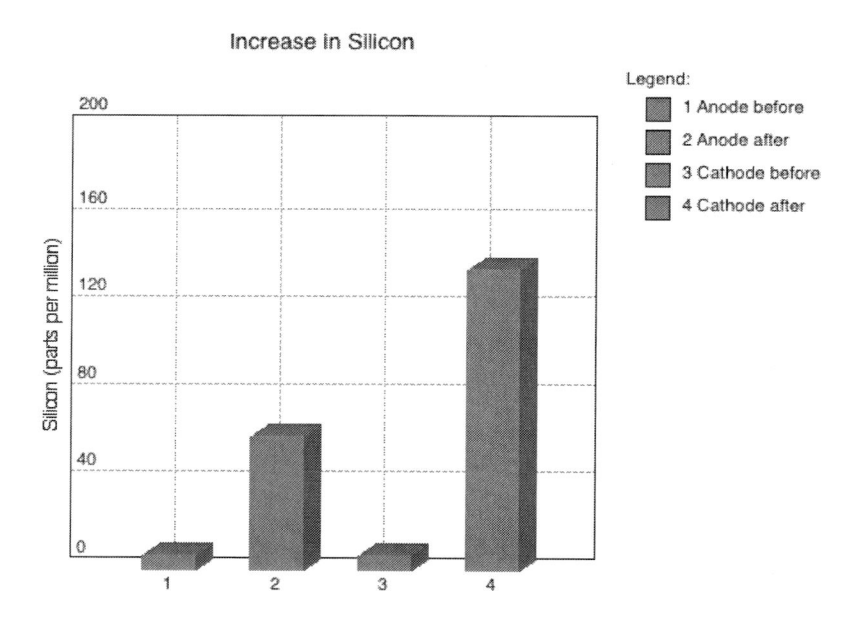

Levels of silicon in anode and cathode before and after the test.

ANOTHER SILICON PATHWAY

As part of the same lab session on Dec. 30, the QR team performed another test that may hint at a different pathway to silicon. In this test, inserts of lithium and sodium were pressed into pure copper electrodes.

Anode and cathode were fabricated out of a pure copper rod. Sulfur test material was placed between the anode and cathode and the tube pumped down to vacuum. Anode and cathode were electrified, producing a glow discharge. After several minutes, the power was disconnected, the tube returned to atmosphere, and the test materials allowed to cool.

Prior to the test, I had predicted that silicon could be formed through the low energy reaction of lithium and sodium, according to the following formula:

$_7Li + {_{23}Na}$ (with oxygen catalyst) → $_{29}Si$
Lithium-7 + sodium-23 → silicon-29

As expected, silicon appeared in the test samples, as did cobalt (Co), which I predicted would appear as the result of a low energy reaction between sodium and sulfur:

$_{23}Na + {_{36}S}$ (with oxygen catalyst) → $_{59}Co$
Sodium-23 + sulfur-36 → cobalt-59

Cobalt could also be the product of a reaction between an atom of sodium-23 and two atoms of oxygen-18 (either from the pure oxygen fill, or through the low energy fission of one atom of sulfur into two atoms of oxygen:

$_{36}S$ (with oxygen catalyst) → $_{18}O + {_{18}O}$

$_{23}Na + 2({_{18}O})$ → $_{59}Co$

Silicon and cobalt appeared in the test samples with the following values in parts per million (NHML File 25985):

	Anode	Cathode	Sulfur Residue
Si	190	45	340
Co	5	4	1

According to the Certificates of Analysis from Alfa Aesar, the lithium rod contained silicon at 4ppm, the sulfur pieces <1 ppm, and the copper electrodes <0.005 ppm. Moreover, cobalt was present only in the copper electrodes but in miniscule amounts, <0.0005 ppm. These quantities are far below the quantities found in the test materials. We should note, however, that silicon is a constituent of the quartz vacuum tube. Thus, in order to more accurately control for silicon, these experiments would need to be conducted in a vacuum chamber made of material in which silicon is not a component.

IRON

The presence of iron in treated graphite may help explain the magnetic properties consistently seen in treated samples. Evidence suggests this magnetic effect may be permanent, having been observed in graphite powder more than a year after the carbon arc test. In the open-air test cited above (NHML File 24067) iron was found at 4700 ppm. It also appeared in the Dec. 30 vacuum test, even though it was not listed in the CofA for the graphite rod. (Iron was listed in the CofA for the sulfur at less than 1 ppm.) ICP analysis on the Dec. 30 graphite samples revealed iron at 8 ppm on the anode and 5 ppm on the cathode (see chart "Increase in Iron.")

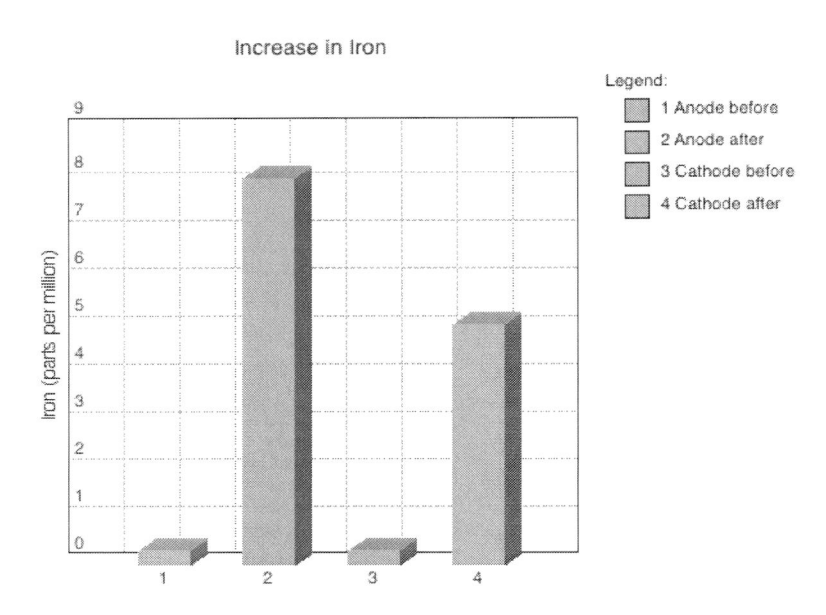

Increase in Iron

Legend:
1 Anode before
2 Anode after
3 Cathode before
4 Cathode after

Levels of iron in anode and cathode before and after the test.

The formation of iron may be an example of the phenomenon known as *proton emission,* a type of radioactive decay in which one or more protons are ejected from a proton-rich nucleus. In Quantum Conversion Theory, iron is formed when two atoms of carbon react with two atoms of oxygen:

$$2(_{12}C + _{16}O) \rightarrow {_{56}}Fe \text{ (minus 2 protons)}$$

In this reaction, the transmuted nucleus has two too many protons (28) and simultaneously ejects the two extra protons to restore equilibrium (28 − 2 → 26Fe). The protons tunnel out of the nucleus in a finite time, in a process known as *quantum tunneling.* To date, more than twenty-five isotopes have been found to exhibit proton emission; proton emitters are not seen in naturally occurring isotopes but are produced via nuclear reactions such as those taking place during the carbon arcing process. The simultaneous emission of two protons (double proton decay) was not observed until 2002 when it was found in an isotope of iron.

Other magnetic isotopes can be produced in this reaction. Nickel has appeared consistently in carbon arc experiments, as has cobalt. Nickel was found in the treated anode at 6 ppm; cobalt in the treated cathode at 2 ppm, although neither was listed in the CofA.

It is possible that nickel (Ni-58) is formed by the reaction of two atoms of carbon-12 with two atoms of oxgyen-17, and by the reaction of two atoms of carbon-12 with two atoms of oxygen-18 (Ni-60.) Cobalt-59 may be produced when atoms of nickel-58 or nickel-60 shed one proton and one or more electrons or neutrons through quantum tunneling. More research is needed to determine the pathways leading to the appearance of these elements.

CARBON NITROGEN REACTIONS

The vacuum test conducted on Dec. 30 was consistent with previous open-air tests in that the same reactions between carbon and oxygen were observed. However, since the test was conducted under vacuum, with pure oxygen backfill, reactions between carbon and nitrogen were non-existent.

For example, aluminum appears consistently in open-air studies but was not found in the vacuum test. In the open-air test cited above (NHML File 24067), aluminum was found in the treated graphite at 7800 ppm. The Quantum Conversion formula for the low energy production of aluminum is as follows:

$$_{12}C + _{15}N \rightarrow _{27}Al$$
Carbon-12 + nitrogen-15 → aluminum-27

Silicon (Si-30) can also react with nitrogen (N-15) to produce scandium (Sc-45), an element that has consistently appeared in open-air graphite studies. Perhaps another round of vacuum studies can be designed in which nitrogen is substituted for oxygen as the fill gas. These studies would test the hypothesis that, in addition to reacting with oxygen, carbon reacts with nitrogen to form new elements.

CONCLUSION

As stated in the introduction, the carbon arc studies conducted under vacuum yielded results consistent with those conducted in the open air, with several important differences, both in the experimental inputs and in the results, as summarized in the table.

Comparison of Carbon Arc Study Methods

Open Air	Vacuum
Greater power input (36-48 Volts DC)	Smaller power input (Approx. 4 amps DC)
Higher temperatures (Graphite electrodes glow white hot)	Lower temperatures (Little or no electrode glow)
Higher pressure (1 atmosphere = 760 torr)	Lower pressure (Approx. 3.5 torr)
Graphite electrodes plus graphite powder (Greater carbon reactivity)	Graphite electrodes only (Less carbon reactivity)
O^2 and N^2 as catalyst gases	O^2 as single catalyst gas (plus solid sulfur)

Wider range of LENR (C + O and C + N reactions)	Narrower range of LENR (C + O reactions only)
Higher volume of converted elements (Mg, Si, Fe, Co, Ni, etc.)	Lower volume of converted elements
Samples test positive for magnetic properties	Magnetic activity untested

Apparently, the stronger electric arc inputs present in the open-air tests, in combination with higher temperatures, the use of fine mesh graphite powders, and the availability of oxygen and nitrogen in the atmosphere, produced a wider array of low energy nuclear reactions with much higher rates of quantum conversion than a parallel test conducted under vacuum. More research utilizing both methods is of course needed to test this hypothesis and confirm or refute these results.

Source: Edward Esko, "Carbon Arc Under Vacuum," *Infinite Energy* Issue 90, March/April 2010.

7. Appearance of Potassium in a Li-S Matrix

ABSTRACT/BACKGROUND

In a study conducted at Quantum Rabbit (QR) lab in Nashua, New Hampshire, USA on February 29, 2008, a lithium test sample was found to contain potassium at 0.14%, a substantial increase over the 36ppm value for potassium listed in the CofA for the pure lithium rods used to create the sample. However, this result was inconclusive due to the fact that the Pyrex (borosilicate) glass used in the vacuum tube contained sodium and potassium oxides at up to 4.2% (Source: M & M Glassblowing, Nashua, NH.) QR researchers were attempting to prove the low energy nuclear reaction: $7Li + 2(16O) \rightarrow 39K$ (lithium-7 reacts with two atoms of oxygen-16 to form an atom of potassium-39.)

In subsequent studies conducted at the Nashua lab and at the QR lab in Owls Head, Maine, on December 30, 2008 and July 30, 2009 respectively, a significant trace of potassium was found on copper electrodes and in lithium/sulfur test material. The amount of potassium detected by ICP (Inductively Coupled Plasma Atomic Spectroscopy) was significantly greater than the amount in the copper electrodes, lithium, or sulfur prior to the test, or the 0.6 ppm present in the quartz used to form the central reaction zone of the glass vacuum tubes.

In order to control for potassium, tubes with a 50 mm center section made of quartz were used. The quartz contained a minute trace of potassium, not enough we believed, to influence the outcome of the test. The quartz tube was used in both lab sessions. The tests were conducted by QR researchers Alex Jack, Woody Johnson, and me, with technical assistance from Bill Zebuhr.

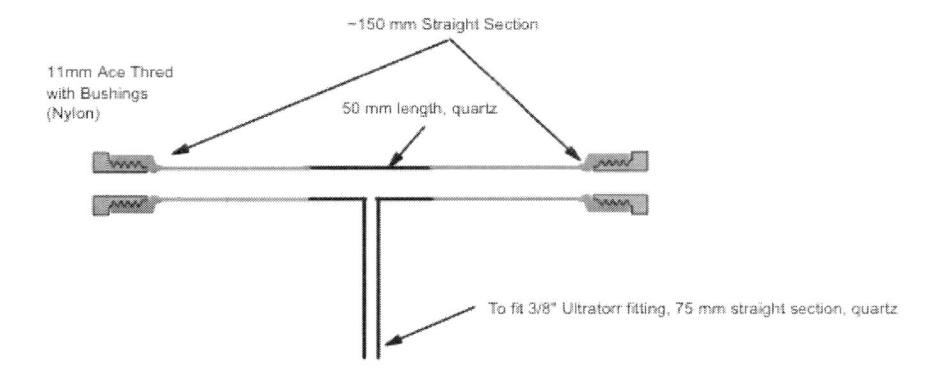

Design of the QR vacuum tube.

METHOD

The same method was employed in the Dec. 30 and July 30 tests. A high purity copper cathode (from Alfa Aesar) was inserted in the QR vacuum tube. The copper was labeled Puratronic® and composed of 99.999% pure copper (metals basis). A small piece of lithium was cut from a fresh lithium rod and placed in the middle section of the tube, together with pure sulfur powder. Like the copper electrodes, the sulfur was labeled Puratronic® and certified as being 99.999% pure.

After placing the sulfur in the tube, the copper anode was inserted at the opposite end and the electrodes connected to the power supply. Both anode and cathode were in contact with the sulfur powder. The tube was pumped down to vacuum and back-filled with oxygen to about 3.5 torr. Electricity was passed through the electrodes, producing a glow discharge. A portion of the lithium and sulfur had apparently vaporized. After several minutes, the electricity was turned off, the vacuum pumps disconnected, and the samples allowed to cool.

RESULTS

In both tests, the electrodes were removed from the tube and packaged for shipping. The tube contained a blackened residue and was also sent to the outside lab, New Hampshire Materials Laboratory, for ICP analysis.

The ICP results for the Dec. 30 test came back on January 9, 2009 (NHML File 25985). The results for July 30 came back on August 14, 2009 (NHML File 26657). Both appeared to confirm our prediction about the appearance of potassium following a low energy nuclear reaction between lithium and sulfur. Here are the differences in the amount of potassium (K) detected in the test materials before and after the experiment. All values are in parts per million (ppm).

	K Before (CofA)	K After Dec. 30	K After July 30
Copper anode	<0.005*	230	638
Copper cathode	<0.005*	73	85
Sulfur	None listed**	38	27
Lithium	36***	See above values for S; noted as "S-Li Residue" in analysis report	

*Alfa Aesar Stock 10156/Lot L08R011
**Alfa Aesar Stock 10755/Lot D03S015
***Alfa Aesar Stock 10773/Lot E31S039

When we add the amount of potassium found in the test materials prior to the experiment, we obtain the following total:

Copper electrodes (2)	<0.010*
Sulfur	0
Lithium	36
Quartz tube material	0.6
Total K	<36.610

*Parts per million

According to the Certificates of Analysis, each copper electrode contained less than 0.005 ppm potassium. There was no potassium listed in the CofA for the sulfur pieces, while the lithium rods contained 36 ppm. When added to the 0.6 ppm potassium contained in the quartz tube, the total equals less than 36.610 pmm potassium. The amount in the anode after the Dec. 30 test equaled 230 ppm, a greater than six-fold increase. On July 30, the amount of potassium on the anode jumped to 638 ppm, a more than seventeen-fold increase over the total amount in the various materials before the test. The increase is greater if we consider the total amount of potassium found in all the samples, including cathode and residue, after the test.

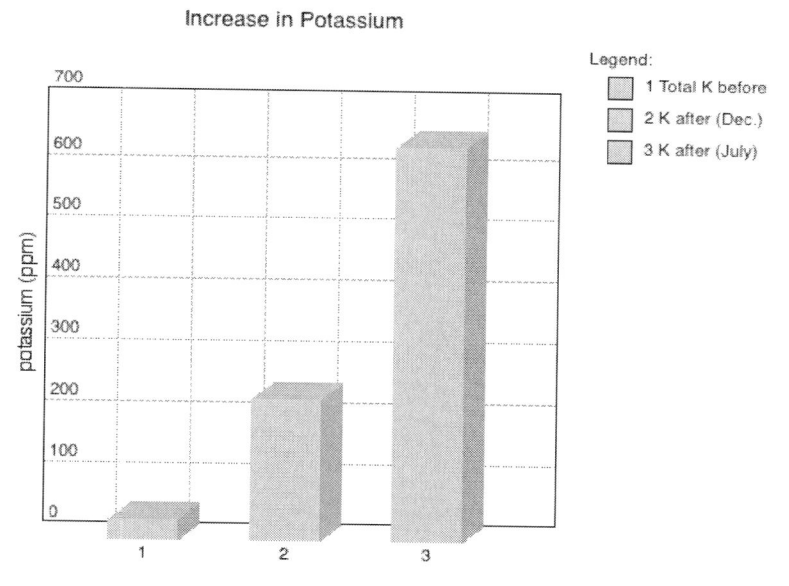

Increase in potassium (as measured on the copper anode) in the Dec. 30 test (2) and the July 30 test (3).

CONCLUSION

The potassium results of Dec. 30, 2008 and July 30, 2009 have helped confirm the predictive power of the quantum conversion theory of low energy transmutation. As predicted prior to both tests, during the experiment, atoms of lithium may have reacted with atoms of sulfur to form potassium:

$_7$Li + $_{32}$S (with oxygen catalyst) → $_{39}$K
Lithium-7 + sulfur-32 → potassium-39

Placing lithium in contact with sulfur, pumping down to vacuum, admitting oxygen as a fill gas, and electrifying the tube may have caused the electrical repulsive force between the lithium and sulfur nuclei to become temporarily neutralized. This, in turn, may have allowed the omnipresent centripetal Casimir force to cause a lithium-7 nucleus to react with a sulfur-32 nucleus to form the heavier nucleus of potassium-39. Moreover, this result may confirm the validity of the earlier formula and attempt to create potassium through a low energy nuclear reaction between lithium and oxygen. Sulfur-32 may well be formed by the low energy nuclear reaction between two atoms of oxygen-16 or $_{16}$O + $_{16}$O → $_{32}$S. In this view, sulfur is a crystallized form of two atoms of oxygen. Given the proper conditions, it may be possible to form potassium from the reaction of an atom of lithium with two atoms of oxygen.

We extend our thanks to the New Energy Foundation (NEF) whose sponsorship and technical assistance made the July 30 test at Owls Head possible.

Source: Edward Esko, "Appearance of Potassium in a Li-S Matrix," *Infinite Energy* Issue 91, May/June 2010.

8. Anomalous Metals in Electrified Vacuum

ABSTRACT
In studies funded by the New Energy Foundation (NEF) and conducted on July 30, 2009 at Quantum Rabbit (QR) lab in Owls Head, Maine, USA, researchers performed several vacuum discharge tests utilizing a copper anode—into which a pure lead insert had been pressed—a copper cathode, pure lithium test material, and a pure sulfur catalyst. Pure oxygen was added to the vacuum tube as a fill gas. Upon analysis by an independent lab, test samples were found to include the anomalous presence of germanium (Ge) at up to 3196 ppm; potassium (K) at up to 750 ppm; and gold (Au) at up to 174 ppm. Although contamination cannot be definitively ruled out as the source of these anomalies, the possibility of low energy nuclear reactions is also a factor to be considered.

BACKGROUND
Although not accepted by mainstream science, investigators around the globe have reported on the possibility of low energy nuclear reactions resulting in apparent element transmutations. These findings have been reported at international conferences, in books, in periodicals such as *Infinite Energy*, and online. In vacuum and carbon arc studies conducted between 2005 and 2009 by Quantum Rabbit LLC and published in *Infinite Energy*, independent analysis of test samples has documented the anomalous appearance of magnesium, silicon, chromium, iron, cobalt, nickel, copper, germanium, strontium, palladium, and tin on electrodes and in test materials. Although contamination of test samples cannot be ruled out, these studies are nevertheless suggestive of low energy nuclear reactions.

THE VACUUM TUBE

The July 30 experiments utilized the special vacuum tube designed by me and employed in previous metal vapor tests. The borosilicate glass tube was 150 mm long, with a 50 mm quartz midsection. A 3/8-inch diameter quartz straight section, perpendicular to the tube and 75 mm in length, connected the midsection with the vacuum manifold, and served as the entrance for a pure oxygen backfill.

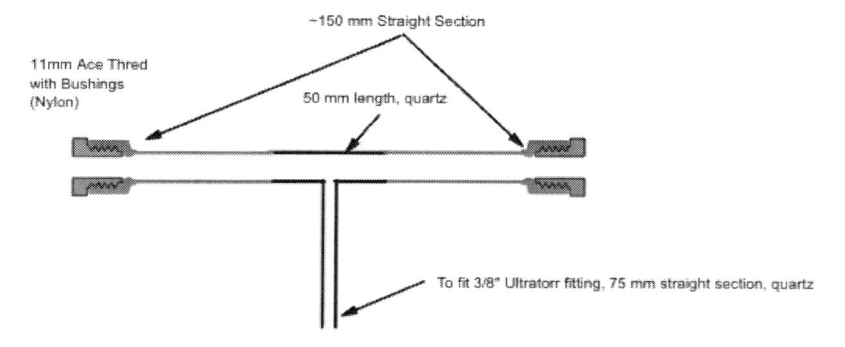

Vacuum tube used in the July 30 test.

Wayne Martin of M & M Glassblowing in Nashua, NH fabricated the vacuum tubes with design assistance from Steve Hansen founder of the Bell Jar and Diverse Arts in Owls Head, ME. The typical trace element composition of the quartz midsection of the tube, the section in which the main reaction took place, is as follows (ppm by weight):

Typical trace element composition of quartz tube material.*

Al	14	Fe	0.2	Sb	<0.008
As	<0.002	K	0,.6	Ti	1.1
B	<0.2	Li	0.6	Zr	0.8
Ca	0.4	Mg	0.1	*OH-	<5
Cd	<0.01	Mn	<0.06		
Cr	<0.05	Na	0.7		
Cu	<0.05	Ni	<0.1		
Fe	0.2	P	<0.2		

*Analysis provided by Wayne Martin, M & M Glassblowing

*May contain a higher amount of surface hydroxyl (OH) ions, but the values represent a bulk average for the total wall thickness.

THE COPPER ELECTRODES

A high purity copper (Cu) rod from Alfa Aesar was cut into two and used to create the anode and cathode. The Alfa Aesar catalogue stated that the Puratronic® (high purity research chemicals and materials) rod was composed of 99.999% pure copper. The Certificate of Analysis (CofA), certified by Alfa Aesar Quality Control, is shown below.

Certificate of Analysis
Copper rod, 9.5mm (0.4in) dia, Puratronic®, 99.999% (metals basis)
Stock Number: 10156
Lot Number: LO8R011

Analysis

Ag	<0.01	Al	0.013	As	<0.002	Au	<0.5
B	<0.0005	Ba	<0.0005	Be	<0.001	Bi	0.0025
Br	<0.005	Ca	0.011	Cd	<0.05	Ce	<0.001
Cl	<0.001	Co	<0.0005	Cr	0.002	Cs	<0.005
F	<0.002	Fe	0.0033	Ga	<0.005	Ge	<0.02
Hf	<0.001	Hg	<0.01	I	<0.002	In	<0.05
Ir	<0.001	K	<0.005	La	<0.01	Li	<0.001
Mg	<0.001	Mn	<0.001	Mo	<0.002	Na	<0.001
Nb	<0.0005	Nd	<0.005	Ni	<0.005	O	<1
Os	<0.001	P	<0.002	Pb	<0.002	Pd	<0.002
Pt	<0.001	Rb	<0.001	Re	<0.001	Ru	<0.005
S	0.055	Sb	<0.005	Sc	<0.0002	Se	0.015
Su	<0.005	Sn	<0.01	Sr	<0.0001	Ta	<5
Te	<0.05	Th	<0.0001	Ti	0.0063	Tl	<0.001
U	<0.0002	V	<0.0002	W	<0.002	Y	<0.0002
Zn	<0.05	Zr	<0.0005				

Values given in ppm unless otherwise noted
Analysis method: GDMS
Certified by Paul V. Connolly, Quality Control, Alfa Aesar®

THE LEAD INSERT

We also ordered a high purity lead (Pb) slug (99.9999% metals basis), and used the lead slug to create a lead insert pressed into a 0.27-inch diameter cavity in the center of the copper anode. Below is the CofA sent by the supplier. **Note:** In handling the lead material, all of the precautionary safety and disposal guidelines set forth in the Materials Safety Data Sheet (MSDS) provided by the supplier were observed.

Certificate of Analysis
Lead slug, 6.35mm (0.25in) dia x 6.35mm (0.25inc) length, Puratronic®, 99.999% (metals basis)

Stock Number: 43415
Lot Number: H18T011

Typical Analysis

Sb	1 ppm
As	1 ppm
Bi	0.2 ppm
Cu	1 ppm
Ag	1 ppm
Tl	2 ppm
Sn	1 ppm
Fe	0.2 ppm
Ca	0.1 ppm
Mg	0.3 ppm

Certified by Paul V. Connolly, Quality Control
Alfa Aesar®

THE LITHIUM TEST MATERIAL

The lithium used in the tests was cut from a pure lithium rod ordered from Alfa Aesar. The rod was listed as 99.9% pure as we see from the following analysis.

Certificate of Analysis
Lithium rod, 12.7mm (0.5in) diax approximately 165mm
(6.50in) long,
99.9% (metals basis)

Stock Number: 10773
Lot Number: E31S039

Analysis

Lithium	100.0%
Sodium	31
Calcium	17
Iron	1
Silicon	4
Chlorine	16
Nitrogen	102
Potassium	36

Values given in ppm unless otherwise noted
Certified by Paul V. Connolly, Quality Control
Alfa Aesar®

THE SULFUR TEST MATERIAL

High purity sulfur, in the form of sulfur pieces, was also ordered from Alfa Aesar and used in the test. The C of A lists the sulfur as Puratronic® 99.999% pure.

Certificate of Analysis
Sulfur pieces, Puratronic®, 99.999% (metals basis)

Stock Number: 10755
Lot Number: D03S015

Analysis

Aluminum	1	Arsenic	<0.2
Bismuth	<0.5	Cadmium	<0.5
Calcium	0.5	Chromium	<0.5

Copper	<0.5	Iron	<1
Lead	<1	Magnesium	0.2
Manganese	0.3	Nickel	<1
Silicon	<1	Silver	<0.2
Sodium	<1	Tin	<0.5
Zinc	<1		

Values given in ppm unless otherwise noted
Certified by (Signature not legible) Quality Control Alfa Aesar®

THE EXPERIMENT

I designed the experiment as follows: first, the copper cathode was inserted in one end of the tube. Small granules of sulfur (S) were placed in the quartz midsection, together with pieces of lithium (Li) cut from the lithium rod. The sulfur, listed in the Alfa Aesar catalogue as "sulfur pieces," was actually a coarse yellow powder that was difficult to maneuver into the tube. However, after persistent attempts, we succeeded in placing a sufficient quantity in the tube. Like the copper electrodes, the sulfur certified by Alfa Aesar as being 99.999% pure (metals basis.) The lithium pieces were also from Alfa Aesar and certified as 99.9% pure (metals basis).

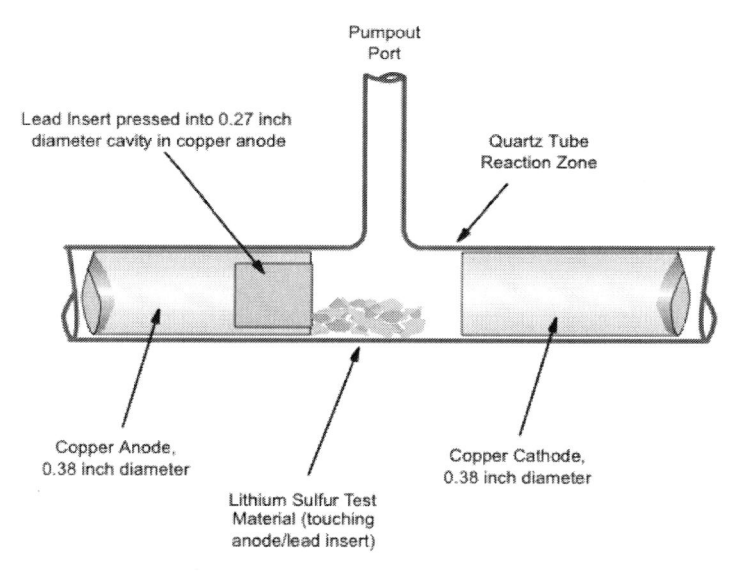

Vacuum tube and electrode configuration used in the July 30 tests.

After placing the sulfur and lithium in the tube, the lead-tipped anode was inserted at the opposite end, and the electrodes connected to the power supply. The lead-tipped anode was in contact with the lithium/sulfur powder. The tube was pumped down to vacuum and backfilled with pure oxygen to approximately 3.5 torr. Electricity was passed through the electrodes, producing an electric arc (approximately 4 amps) and glow discharge. After several minutes, the electricity was turned off, the vacuum pumps disconnected, and the sample allowed to cool.

QR researchers Woody Johnson (left) and Alex Jack (right) with Bill Zebuhr at Owls Head July 30, 2009.

TEST RESULTS

The experiment was conducted three times on July 30, with minor variations between experiments. Three samples were retrieved from each of the three tubes: the lead-tipped anode, the copper cathode, and sulfur-lithium residue from the center of the tube. The anode and cathode tips had undergone noticeable changes during the experiment, with the presence of an undetermined residue on their surface. The S-Li residue had also changed. All nine samples were packaged and sent to New Hampshire Materials Laboratory for ICP analysis. ICP results came back on August 14, 2009 (NHML File Number 26657). The test report is reproduced below.

New Hampshire
MATERIALS
Laboratory, Inc.

Test Report

August 14, 2009

Mr. Edward Esko
Quantum Rabbit LLC

File Number: 26657

Overview:
Samples Received: (3) Lots of residue samples as noted below
Work Requested: Chemical analysis by ICP
Sample Disposition: Return remains to client

Analysis Results:

Test 1	Lead Anode	Copper Cathode	S-Li Residue
Germanium	1018 ppm	2190 ppm	388 ppm
Potassium	638 ppm	85 ppm	27 ppm
Gold	162 ppm	11 ppm	1 ppm
Sample weight (gm)	0.0157	0.0494	0.2800

Test 2	Lead Anode	Copper Cathode	S-Li Residue
Germanium	102 ppm	119 ppm	31 ppm
Potassium	16 ppm	31 ppm	6 ppm
Gold	5 ppm	<1 ppm	<1 ppm
Sample weight (gm)	0.433	0.1047	0.7861

Test 3	Lead Anode	Copper Cathode	S-Li Residue
Germanium	35 ppm	2 ppm	7 ppm
Potassium	17 ppm	10 ppm	1 ppm
Gold	7 ppm	<1 ppm	<1 ppm
Sample weight (gm)	0.0290	0.1283	0.6961

Prepared by:
Timothy M. Kenney
Director of Laboratory Services

ANALYSIS

Where did the anomalous metals come from? According to the most common interpretation, which I refer to as the contamination theory, the anomalous elements were present in the experiment at the beginning and were not detected until the test materials were analyzed. That is a perfectly reasonable assumption given the current model of physics. It is helpful in considering this possibility to review the analysis data for the materials used in the experiment to see if and where the anomalous elements show up. (The one input for which we do not have specific data is the pure scientific grade oxygen supplied by Spec Gas, Inc. and used as backfill in the tube. Spec Gas certifies this scientific grade oxygen as 99.99999% pure. See Specgas.com for more information.)

Inputs	Germanium	Potassium	Gold
Quartz Tube Material	ND*	0.6 ppm	ND
Copper Electrodes	<0.02 ppm	<0.005 ppm	<0.5 ppm
Lead Insert	ND	ND	ND
Lithium Pieces	ND	36 ppm	ND
Sulfur Pieces	ND	ND	ND

*Element sought but not detected.

Outside contamination is certainly a possibility, for example, from the air in which the samples were exposed or in some manner during the handling of the materials prior to and following the tests. Also it may be that the data in the Certificates of Analysis (CofA) are unreliable, representing generic batches of product rather than the lots out of which the actual test materials were culled. For example, even though the CofA for the lead slug shows no presence of gold, certain lead alloys, such as Novodneprite ($AuPb_3$) named after the region in Kazakhstan where it is mined, occur in combination with gold. This mineral is made up of 75.94% lead and 24.06% gold. Although remote, it is possible that the actual lead sample used in the experiment came from the Novodneprovskoe Deposit and was contaminated with minute traces of gold; contamination that went undetected by the supplier but which appear in ICP analysis following the experiment.

I invite readers to review the data and suggest alternate pathways for contamination. If contamination can be proven without a doubt as the source of the anomalous metals, then we have no need for further research in this area. I challenge investigators with the correct lab apparatus and controls to prove or disprove this possibility. At the same time, ideas pointing to the possible source of contamination could lead future investigators to design carefully controlled experiments yielding more definitive results.

The other possibility can be referred to as the concentration theory. In this hypothesis, the trace amounts of the anomalous elements present in the test materials are somehow "concentrated" during the experimental process so that they appear in higher concentrations in the final analysis. So for example, potassium appears in the lithium test material at 36 ppm, the quartz tube material at 0.6 ppm, and the copper electrodes at <0.005 ppm. The electrification and heating of the test materials may concentrate the already existing potassium in the test sample, especially in the quartz used for the part of the tube that came into contact with test material. However, whether an experiment such as this is capable of concentrating potassium from the combined total of 36.605 ppm documented in the materials prior to the experiment to the 638 ppm recorded after the experiment remains to be seen.

Since germanium is listed as occurring in the copper rod at <0.02 ppm, it is difficult to see how the concentration theory would apply, unless, once again, the Certificate of Analysis provided by the supplier is unreliable. Please note that the amount of germanium detected on copper cathode following Test 1 is 2190 ppm. The only source of gold shown in the Certificates is the <0.5 ppm listed for the copper rods used to make the electrodes. Again, the question whether the experimental process is capable of increasing the concentration of Au from <0.5 ppm as reported in the copper prior to experiment to the 162 ppm detected after the experiment awaits investigation.

The third possibility, one that is not accepted by modern science, is that at least three low energy nuclear reactions took place during the experiment. According to the standard model of physics, element transmutation cannot occur under the conditions that took place in these experiments.

Writing in *Infinite Energy* ("Was Transmutation Observed at the Quantum Rabbit Laboratory," Issue 92), physicist Matthias Grabiak states: "The idea of nuclear re-actions between heavy nuclei as supposedly observed in the Quantum Rabbit experiments is like suggesting that an ant is capable of pulling a freight train. The electric repulsion becomes so forbiddingly strong that such reactions at low energies would be thoroughly ruled out by the known laws of physics. Conventional physics tells us that it requires pressures and temperatures comparable to those of the interior of the sun to overcome the Coulomb barrier even for the very lightest elements in order to achieve fusion. But, for fusing heavier elements, only cataclysmic scenarios like supernova explosions will suffice."

Placing the sulfur-lithium test material in contact with the lead anode, pumping down to vacuum, admitting oxygen as a catalyst, and electrifying the tube may have caused the electrical repulsive force existing between two plus charged nuclei—lithium and sulfur—to become temporarily neutralized. Such temporary neutralization may have allowed the omnipresent centripetal Casimir force to cause a nucleus of lithium-7 and a nucleus of sulfur-32 to react and form potassium-39. As the fusion process took place, a simultaneous fission reaction occurred, in which the highly saturated and unstable lead anode fractured and surrendered atoms of lithium ($_{204}Pb - {_7}Li \rightarrow {_{197}}Au$), or lead-204 minus lithium-7 $\rightarrow$ gold-197. The equation can also be written as:

$$_{204}Pb \rightarrow {_7}Li + {_{197}}Au$$

Meanwhile, in another region of the periodic table, lithium may have reacted with copper to form germanium ($_{63}Cu + {_7}Li \rightarrow {_{70}}Ge$) All in all an intriguing set of possibilities.

Source: Edward Esko, Anomalous Metals in Electrified Vacuum, *Infinite Energy* Issue 99, Sept./Oct. 2011.

9. In Search of the Platinum Group Metals

ABSTRACT

In studies sponsored by the New Energy Foundation (NEF) and conducted at Quantum Rabbit (QR) lab in Owls Head, Maine, on the scenic Penobscot Bay, on December 17, 2009, a QR research team including Alex Jack, Woody Johnson, and me, assisted by Bill Zebuhr and vacuum consultant Steve Hansen, conducted vacuum discharge tests utilizing pure graphite rods in combination with strontium and sulfur test materials. The goal of these studies was to see if we could trigger low energy nuclear reactions in which strontium would react with carbon and with atmospheric fill gases to form ruthenium and other platinum group metals.

BACKGROUND

Ruthenium is one of the rarest elements on earth. It is the first of the precious metals, and is part of the platinum group of metals, occurring naturally in ores in which platinum is found. Earlier studies at QR labs suggested the possibility of inducing low energy nuclear reactions under relatively low temperatures and pressures, and with relatively small inputs of energy. In previous studies, ICP (Inductively Coupled Plasma Atomic Emission Spectroscopy) analysis of Quantum Rabbit test samples had noted the anomalous appearance of copper, palladium, strontium, chromium, germanium, and tin on electrodes and test materials used in QR vacuum studies.

These results point to the possibility of low energy nuclear reactions, in which lighter elements react, under relatively low temperature, pressure, and energy, and create heavier elements.

METHOD

Like previous QR vacuum studies, the Dec. 17 tests took place in a specially designed vacuum tube. The borosilicate tube was 150 mm long, with a 60 mm quartz midsection. A 3/8-inch diameter quartz straight section, perpendicular to the tube and 75 mm in length, connected the quartz reaction zone with the vacuum manifold, and served as the entrance for backfill from the surrounding atmosphere.

We had ordered high purity graphite rods from Alfa Aesar and used the rods to create the anode and cathode. The graphite electrodes were inserted into aluminum electrodes, 0.38-inch in diameter, designed to be flush with the inner wall of the tube so as to seal the tube to maintain vacuum. The aluminum electrodes were positioned outside the quartz reaction zone so as not to interfere with reactions taking place in the center of the tube.

The graphite rods were ultra high purity, 99.9995% pure (metals basis). Their CofA is as follows (Stock 14754/Lot F27S027):

Graphite electrode, counter-flat top, 3.05mm dia, 38.10 mm long, 99.9995% (metals basis)

Al	ND	Fe	ND	Ni	ND	W	ND
B	ND	Pb	ND	Si	ND	V	ND
Ca	ND	Mg	ND	Ag	ND	Zn	ND
Cr	ND	Mn	ND	Sn	ND	Zr	ND
Cu	ND	Mo	ND	Ti	ND		

Values given in ppm unless otherwise noted
ND: Not detected

First, the graphite cathode was secured in the tube, and a catalyst of pure sulfur placed in the quartz midsection. The sulfur, listed in the Alfa Aesar catalogue as "sulfur pieces," was actually a coarse powder. It was labeled as Puratronic® (Stock 10755/Lot D03S015) and certified as being 99.999% pure (metals basis.) Like the graphite rods, the sulfur powder contained no trace of ruthenium or any other platinum group metal.

After the sulfur was placed in the tube, a piece of strontium was positioned in the center of the reaction zone. The strontium, listed in the Alfa Aesar catalogue as "strontium granules," had started to oxidize and had a thin grey coating on the surface. The CofA from Alfa Aesar for the strontium granules showed no trace of ruthenium or any other platinum group metal (Stock 35789/Lot G25R040.)

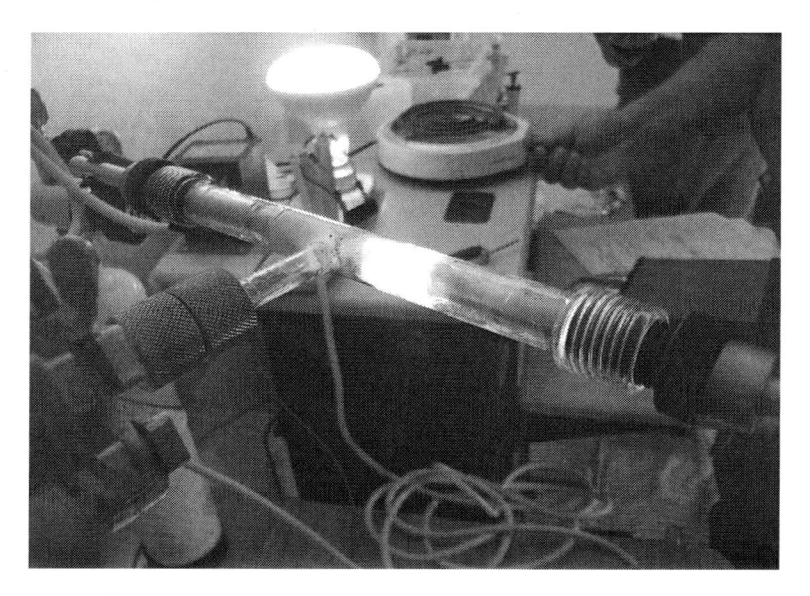

Strontium-sulfur plasma in the QR tube, Dec. 17, 2009.

Following insertion of the sulfur and strontium, the graphite an ode was inserted in the opposite end of the tube, flush with the strontium piece. The electrodes were connected to the power supply. The tube was pumped down to vacuum and backfilled with pure oxygen to approximately 3.5 torr. Direct current (DC)was passed through the electrodes, producing an electric arc and glow discharge with the blue/white characteristic of sulfur. After about two minutes, the current was lowered to a level sufficient to maintain the arc. This was sustained for about five minutes, at which time the current was boosted to the level sufficient to spark an active discharge. This alternating high-low-high-low pulse was maintained for over twenty minutes.

Also, during this period, electrode polarity was reversed, with anode becoming cathode and cathode becoming anode. After the test period, the electricity was turned off, the vacuum pumps disconnected, and the sample allowed to cool.

The test was then repeated, with another set of graphite electrodes and strontium and sulfur test materials. The protocol was essentially the same with several slight differences. In Test 2, the strontium test material was placed in contact with the graphite cathode rather than the anode. Also, during the approximately twenty minute test period, we varied the amount of pressure in the tube, with a low of 3.5 torr and a high of 30 torr. During the period of slightly elevated pressure, we noticed that the strontium vapor glowed brightly with a reddish-violet hue.

Alex Jack (left) and Bill Zebuhr monitor the Dec. 17 test.

RESULTS

Three samples were retrieved from each test: the graphite anode, cathode, and strontium-sulfur residue.

The samples were packaged and sent to New Hampshire Materials Laboratory for ICP analysis. ICP results came back on January 7, 2010. In addition to ruthenium, we asked NHML to scan for other platinum group metals including rhodium, palladium, and platinum. ICP analysis of the test materials came back with the same values for all six samples (NHML File 27109/January 7, 2010):

Element   Test 1 and 2 Graphite Electrodes/Sr Residue

Ruthenium   <26ppm
Palladium   <100ppm
Rhodium   <32ppm
Platinum   <5ppm

Values in parts per million (ppm)

Due to the small size of the samples, ranging from a low of 0.0733 gm to a high of 1.5419 gm, none of the requested elements was found down to the detection limits listed in the report. Below the detection limits it was impossible to determine whether or not the requested element was present. In other words, the results were inconclusive. It was impossible to state for certain whether we did or did not achieve the sought-after low energy transmutations.

ANALYSIS

As mentioned above, we had selected ruthenium as the target metal for the Dec. 17 tests. Our hope was to prove the formula:

$_{88}Sr + _{12}C \rightarrow _{100}Ru$
Strontium-88 + carbon-12 → ruthenium-100

In addition, we had speculated that a variety of other low energy transmutations would take place involving the elements used in the experiment, including nitrogen and oxygen from the open air. These low energy reactions included:

$_{88}Sr + _{15}N \rightarrow _{103}Rh$
Strontium-88 + nitrogen-15 → rhodium-103

$_{88}Sr + _{16}O \rightarrow _{104}Pd$

Strontium-88 + oxygen-16 → palladium-104

Several factors may have influenced the inconclusive result, including:

1. Failure to reach high enough temperature. All of the platinum group metals have very high melting points. It may be necessary to achieve temperatures close to or above these points to facilitate low energy transmutation. The glass vacuum tube used in the experiment may not be able to sustain such high temperatures. A new tube design may need to be employed for this purpose.

2. Oxidation of the strontium test sample. The oxide on the surface of the strontium pieces may have interfered with the availability of pure strontium for a low energy nuclear reaction. Future tests should use fresh strontium with no oxide coating.

Due to the oxidation of the strontium test sample, the QR team was essentially dealing with strontium oxide rather than pure strontium. A brief comparison between the two may shed light on the result achieved in the QR experiment.

	Strontium	Strontium Oxide
Melting Point	777 degrees C	2531 degrees C
Boiling Point	1382 degrees C	>3000 degrees C

The higher vaporization point of strontium oxide versus pure strontium no doubt played a role in the outcome of the test.

3. Use of atmosphere as fill gas rather than pure oxygen. Previous tests in which a definitive result was achieved employed pure oxygen as the fill gas. The use of atmosphere (nitrogen/oxygen) as fill rather than pure oxygen may have impeded potential low energy nuclear reactions.

It is important to note that we have received similar results in past tests, only to see the predicted formulas apparently proven in subsequent studies. In tests conducted on Sept. 30, 2008, I had predicted the appearance of germanium-70 on copper-63 electrodes following a low energy reaction with lithium-7.

The analysis of the Sept. 30 test materials came back inconclusive, with the detection limits reached on all samples. However, in subsequent tests, germanium was found on test samples at up to 416 ppm (on Dec. 30, 2008) and 3196 ppm (on July 30, 2009). See "Appearance of Tin on a Silver Anode," *Infinite Energy* Issue 88. Whether any of the platinum group metals will appear with certainty in repetitions of the above test remains to be seen.

Further research is of course needed to prove or disprove these assertions. Rather than being definitive, Quantum Rabbit research has been *suggestive* of the possibility of low energy nuclear reactions. It is hoped that additional research will expand our knowledge of this phenomenon and open the possibility of a new paradigm of scientific understanding and sustainable technology.

Source: Edward Esko, "In Search of the Platinum Group Metals," *Infinite Energy* Issue 92, July/August 2010.

10. Quantum Tunneling and the Quantum Rabbit Effect

I would like to respond to the article, "Was Transmutation Observed at the Quantum Rabbit Laboratory," *Infinite Energy* Issue 92, July/August 2010, in which Matthias Grabiak states that the Coulomb barrier between positively charged protons is too strong to allow low energy fusion to take place. As he puts it: "Conventional physics tells us that it requires pressures and temperatures comparable to the interior of the sun to overcome the Coulomb barrier even for the very lightest elements to achieve fusion. But, for fusing heavier elements as suggested by the Quantum Rabbit experiments, only cataclysmic scenarios like supernova explosions will suffice."

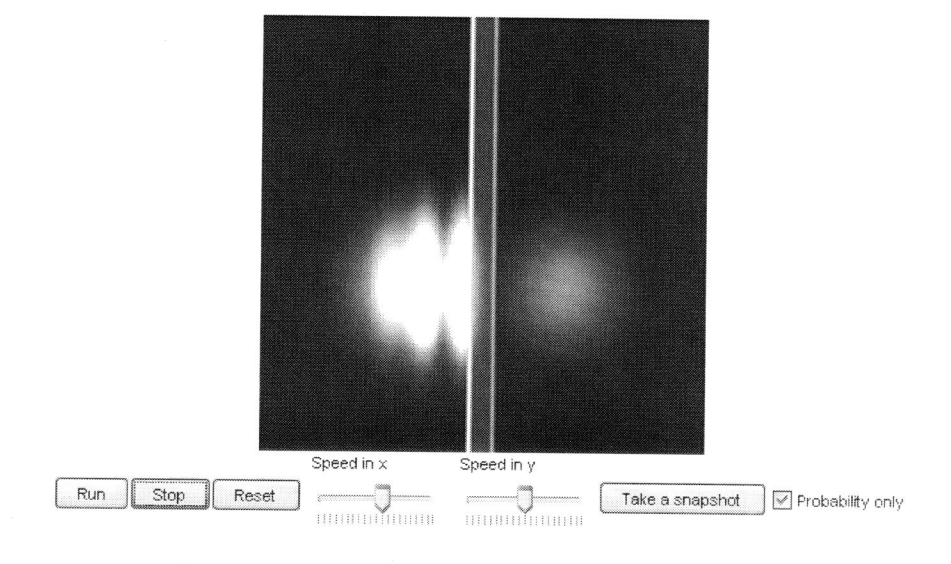

It may be that the phenomenon of quantum tunneling offers an explanation of how the Coulomb barrier can be breached with relatively low inputs of temperature, pressure, and energy and how nuclear transmutation can be achieved under these conditions. In the above illustration we see the reflection and tunneling of an electron wave packet directed at a potential barrier. The bright spot moving to the left is the reflected part of the wave packet. The dim spot moving to the right is the small fraction of the wave packet that tunnels through the classically forbidden barrier.

Applying this phenomenon to the QR experiment in which the possible low energy transmutation of lithium into potassium was noted ("Appearance of Potassium in a Li-S Matrix," *Infinite Energy* Issue 91), the bright spot on the left corresponds to the lithium nuclei used in the test. The potential barrier represents the Coulomb barrier existing between two positively charged nuclei, in this case, those of lithium and sulfur used in the experiment. In this model, a small fraction of the lithium nuclei, which exist in the form of a wave packet (represented by the dim circle on the right), tunnels through the barrier and moves near enough to the sulfur nuclei for the strong force to bind them together to form larger nuclei of potassium ($_7Li + _{32}S \rightarrow _{39}K$). Please note that in my articles I refer to the strong force as the centripetal Casimir force found throughout the universe.

In this model, the nucleus behaves not as a classical particle but as a quantum wave that obeys the same laws of quantum tunneling as an electron wave.

The diagram below offers a schematic version of this possibility. In the experiment described above, the wave to the left of the barrier (vertical line) represents the total number of lithium nuclei approaching the Coulomb barrier in the form of a wave packet. The wave to the right of the barrier in the lower diagram on the next page represents the fraction of lithium nuclei that tunnel through the barrier and enter into a low energy reaction with the sulfur nuclei.

Of course, as I have mentioned in my articles, the Quantum Rabbit experiments are suggestive of low energy transmutation, and not yet definitive. Similar results to those achieved by Quantum Rabbit in laboratories around the world will be needed to establish low energy transmutation as a reality.

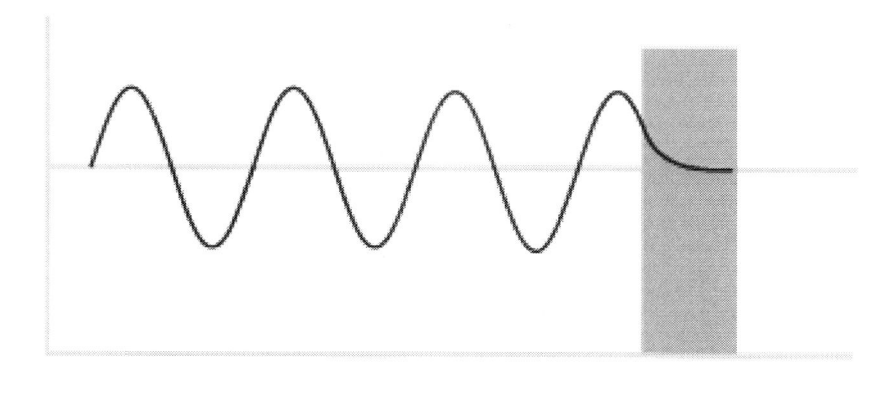

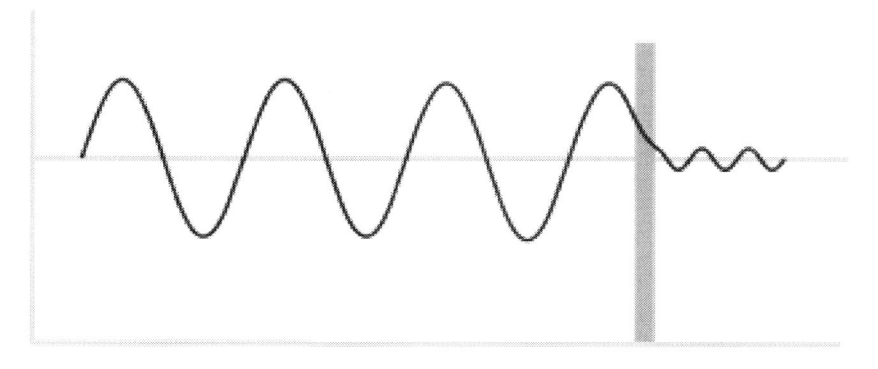

Source: Edward Esko, "Letter," *Infinite Energy* Issue 93, Sept./Oct. 2010.

11. The Possibility of Plutonium Reduction

The issue of what to do with the nuclear waste left over from nuclear reactors and nuclear weapons programs remains a vexing problem in countries around the world. So far, no satisfactory solutions have emerged. Nuclear waste is clearly an impediment to a clean natural environment and poses a threat to world peace.

Environmental, health, and public interest groups including Greenpeace, Physicians for Social Responsibility, Friends of the Earth, and the Natural Resources Defense Council are united in opposition to the reprocessing of nuclear waste, a practice in which radioactive plutonium and uranium are separated from used or "spent" nuclear fuel from nuclear power reactors. The reprocessed waste is then reused as fuel.

In a May 2008 fact sheet entitled "Reprocessing: Dangerous, Dirty and Expensive–Why Extracting Plutonium from Nuclear Reactor Spent Fuel is a Bad Idea," the Union of Concerned Scientists stated that reprocessing would increase the risk of nuclear terrorism:

> Less than 20 pounds of plutonium is needed to make a nuclear weapon. Commercial-scale reprocessing facilities handle so much of this material that it has proven impossible to keep track of it accurately. A U.S. reprocessing program would add to the worldwide stockpile of separated and vulnerable plutonium that sits in storage today, which totaled roughly 250 metric tons as of the end of 2005—enough for some 40,000 nuclear weapons. Reprocessing the U.S. spent fuel generated to date would increase this by more than 500 metric tons.

In a December 9, 2008 letter to then President-elect Obama, a coalition of 150 environmental and health groups pointed out that reprocessing would increase environmental contamination and threaten public health, cost hundreds of billions of dollars of taxpayer funds, and not solve the nuclear waste problem-not even in France:

> Although France reprocesses all its spent nuclear fuel, it is faced with the same difficulties the United States has in siting a permanent geologic repository. The proposed permanent repository site in Bure, France faces overwhelming public opposition, similar to Yucca Mountain in Nevada. In addition, reprocessing has polluted the environment, including the ocean as far away as the Artic Circle, and has created a stockpile of more than 80 metric tons of separated plutonium.

The coalition of public interest groups went on to urge the Obama administration to focus on securing nuclear waste at reactor sites. However well informed, this measure doesn't solve the problem of what to do with existing plutonium reserves. As we move forward, our goal should be not simply to bury radioactive waste in reactor sites, but to explore the possibility of reducing or even eliminating existing stocks of plutonium and other radioactive elements.

THE NATURE OF PLUTONIUM

In *A Guide to the Elements Second Edition* (Oxford University Press) Albert Stwertka describes plutonium (Pu):

> Plutonium [atomic number 94] is the most important of the transuranium elements, all of which follow uranium [atomic number 92] in the periodic table and all of which are artificially made.

In *Nature's Building Blocks: An A-Z Guide to the Elements* (Oxford University Press) John Emsley describes the destructive power of this manmade element, which has no role in nature or in the human body:

By the spring of 1945 several kilograms had been amassed, and the first atomic explosion, using plutonium, took place at Alamogordo, in the desert of New Mexico, on July 16. It detonated 6 kilograms of plutonium and was triggered by using small conventional explosives to force several pieces of plutonium together to give the critical mass necessary for a runaway chain reaction.

The second plutonium explosion was in the form of a bomb, code-named 'Fat Man' which was dropped on the Japanese city of Nagasaki on 9 August 1945. The explosive capacity was equivalent to several thousand tons of TNT, killing about 70,000 citizens and wounding 100,000.

The even more destructive hydrogen bombs are themselves triggered by a plutonium bomb whose explosion generates temperatures high enough to cause hydrogen atoms, in the form of the heavier isotopes, deuterium or tritium, to fuse together. Because of the fission-generated heat needed to generate the fusion reactions, these are classified as thermonuclear weapons. As hydrogen atoms fuse, they emit vastly more energy than an atomic bomb and give explosions equivalent to millions of tons of TNT.

THE PLUTONIUM MINUS SODIUM PATHWAY

In Quantum Conversion Theory, in which elements can be induced to change into each other, with relatively low inputs of energy, it may be possible to produce low energy nuclear fusion reactions, which in turn trigger simultaneous low energy nuclear fission reactions. It may be possible to prompt large, heavy, potentially unstable atoms, such as lead, uranium, or even plutonium, to shed protons and neutrons and convert into lighter more stable elements without destructive consequences. Quantum Rabbit low energy transmutation studies, conducted since 2005 and reported in *Infinite Energy*, suggest such a possibility. Holistic educator, Michio Kushi, suggested the possibility of low energy nuclear fusion/fission reactions as early as the 1960s. The Quantum Conversion approach to the problem of plutonium reduction is simple, direct, and balanced. Atoms of sodium (atomic number 11) are subtracted from atoms of plutonium (atomic number 94) to form atoms of bismuth (atomic number 83). This low energy fission reaction is triggered by a low energy fusion reaction in which sodium-23 reacts with sulfur-34 (with oxygen as a fill gas under vacuum), to form cobalt-27. This reaction serves as a vector that "pulls" atoms of sodium from atoms of plutonium, thus yielding bismuth. Keep in mind that our proposal is to initiate these reactions at low temperature, pressure, and energy, without the release of radioactivity into the environment. On the contrary, our goal is to *reduce* the amount of radioactive material in the environment.

Proposals to store radioactive waste in salt mines are hinting at this reaction. There seems to be an intuitive awareness that salt, or sodium, can somehow neutralize radioactivity. Quantum Rabbit research with other large heavy nuclei suggests the reaction could be achieved in a sealed tube at approximately 3.5 torr with as little as a 4 amp discharge between anode and cathode. Large unstable atoms at the far end of the periodic table may actually be quite easy to fraction into two or more lighter elements. The formula is as follows:

$$Pu \rightarrow Na^* + Bi^* \; (+ \; neutrons)^*$$

*product of low energy fission

In addition to the reduction of plutonium, and the production of bismuth and cobalt, this reaction would most likely produce the discharge or release of neutrons. Plutonium has 19 known isotopes while bismuth has one natural isotope, $_{209}Bi$. The principal isotopes of plutonium, the way they may react to form bismuth, and the possible shedding of neutrons, is presented below:

$$_{238}Pu \rightarrow \; _{23}Na + \; _{209}Bi + 6n^*$$
$$_{239}Pu \rightarrow \; _{23}Na + \; _{209}Bi + 7n$$
$$_{240}Pu \rightarrow \; _{23}Na + \; _{209}Bi + 8n$$
$$_{241}Pu \rightarrow \; _{23}Na + \; _{209}Bi + 9n$$
$$_{242}Pu \rightarrow \; _{23}Na + \; _{209}Bi + 10n$$

*n equals the number of neutrons discharged during the reaction

Quantum Rabbit is not equipped to handle plutonium or conduct the above-suggested research. However, we would be happy to serve as technical or theoretical advisors to established laboratories such as Oak Ridge, Los Alamos, Bhabha Atomic Research Center in India, and others around the world. As stated above, the goal of this research is to secure a clean natural environment for future generations and reduce potential nuclear threats to human health, peace, and well being.

Source: Edward Esko "The Possibility of Plutonium Reduction," *Infinite Energy* Issue 84, March/April 2009.

12. Lessons from Japan's Nuclear Crisis

The nuclear reactors in Fukushima, Japan, the site of multiple meltdowns following the giant earthquake and tsunami on March 11, 2011.

The earthquake and tsunami that occurred on March 11, 2011 have exposed the weakness of Japan's nuclear program. Japan's policy on nuclear energy is misguided and naïve. Prior to March 11, Japan's 54 reactors provided some 30% of the country's total electricity production. There are ambitious plans to build more reactors to increase this share to 41% by 2017 and 50% by 2030.

This policy is moving in the wrong direction. Rather than increasing its dependence on nuclear power to 50% by 2030, Japan should strive to reduce it's dependence by 50% by that date, with the stated goal of eliminating nuclear power before the end of the century, while, at the same time launching an all-out push to develop sustainable technologies such as wind, solar, hydro, geothermal and biothermal.

Simply substituting green technologies for nuclear power will not solve the whole problem, however. There remains the problem of what to do with spent nuclear fuel, which, as many have discovered in this latest crisis, is highly radioactive and highly dangerous.

The worldwide inventory of spent nuclear fuel was 220,000 tons in the year 2000, and is increasing by about 10,000 tons a year. Even with billions of dollars spent on a variety of disposal options, the nuclear industry and governments have not come up with a feasible and sustainable solution.

Most current proposals for dealing with nuclear waste involve burying it in deep underground sites. Whether the storage containers, the store itself, or the surrounding rocks will offer enough protection to stop radioactivity from escaping in the long-term remains unknown.

Japan is currently pursuing a policy of reprocessing nuclear waste at a cost of trillions of yen. Recycling nuclear waste is a distortion of the concept of recycling. Used nuclear fuel has been shipped to the UK and France where it has been reprocessed and returned to Japan as reactor fuel, notably as mixed oxide (MOX) fuel. Labeling this practice "recycling" is a deception designed to convey a "green" veneer to the process.

Rather than "recycling" nuclear waste, Japan should seek to eliminate it, beginning with highly toxic plutonium. The dark history of this destructive manmade element is especially relevant to Japan. The first atomic bomb tested at Alamogordo in 1945 used plutonium to create critical mass necessary for an atomic explosion. This process was repeated on August 9, 1945 with the explosion of a plutonium bomb over the city of Nagasaki.

Rather than investing trillions of yen in additional reprocessing facilities, and essentially not dealing with the problem, those funds should be invested in promising new technologies that offer the possibility of eliminating these toxic materials.

Promising new technologies include using microorganisms to convert toxic waste into harmless elements through the process of biological transmutation. Another promising technology is the laboratory process of cool fusion and cool fission, in which heavy unstable radioactive nuclei are induced to shed mass and convert to more stable non-radioactive elements at low temperatures, pressures, and energies. In this process, suggested by Quantum Rabbit research, plutonium-239 and uranium-235 are converted to bismuth-209 and lead-208 which, although not ideal as end points in the nuclear reduction process, are nevertheless less dangerous than their radioactive counterparts (see "The Possibility of Plutonium Reduction," *IE* Issue 84, March/April 2009.)

We propose that the nuclear industry take the lead in research ways to eliminate nuclear waste. A global "clean the planet" campaign funded by the nuclear industry would help solve the problem. Even If the proposed conversion to non-toxic materials reduces the problem of nuclear waste, because of it's inherent dangers, nuclear power should nevertheless take a back seat to the development of sustainable technologies. Given the events of March 11, a global moratorium on construction of any new nuclear power plants seems like a prudent course of action coupled with a plan to implement nuclear phase-out as green replacement technologies are phased in.

The events of March 11 have brought our choices into sharp focus. Japan and the rest of the world have the choice of whether to continue on the current path, and face the possibility of future accidents and disasters, or to create a safer path toward genuine sustainability.

Source: Edward Esko, "Letter," *Infinite Energy* Issue 98, July/August 2011.

13. Anomalous Metals Part II

ABSTRACT

In a study funded by the New Energy Foundation and conducted at Quantum Rabbit (QR) lab in Owls Head, Maine, on September 27, 2011, independent analysis by inductively coupled plasma spectroscopy (ICP) of test samples revealed the anomalous appearance of potassium (K) at 181 ppm (parts per million) and gold (Au) at 252 ppm. The vacuum discharge test employed a copper cathode, pure lithium and sulfur test material, and a pure copper anode at the center of which a pure lead insert had been pressed. Scientific grade neon was added to the vacuum tube to strike plasma, followed by a catalyst of pure oxygen. Although it is possible that test materials were contaminated, the appearance of these anomalous metals once again raises the possibility of low energy transmutation.

BACKGROUND

The 2011 test followed a series of tests conducted at the QR lab in Owls Head on July 30, 2009, and described in my paper "Anomalous Metals in Electrified Vacuum" published in *Infinite Energy*, Issue 99. In the 2009 tests, a lead insert was pressed into a copper anode. A copper cathode was inserted into one end of the vacuum tube, followed by pure lithium and sulfur test material.

The lead-tipped anode was then inserted, the tube pumped down to vacuum, and pure oxygen admitted to approximately 3.5 torr. An electric arc was struck and a glow discharge with the characteristic color of lithium was produced.

ICP analysis of test samples revealed the anomalous presence of germanium (Ge) at up to 3196 ppm; potassium (K) at up to 750 ppm; and gold (Au) at up to 174 ppm. Once again, aside from possible minute traces listed in the Certificates of Analysis of several test components, none of the anomalous metals was introduced into the experiment.

VACUUM TUBE DESIGN

The 2011 test employed a vacuum tube with a different design than that used in 2009. In the 2009 tests, a vacuum tube made of borosilicate was positioned horizontally atop the vacuum manifold and attached to the manifold by a pump out port. The tube contained a quartz midsection where the reactions took place. The 2011 tube was made entirely of quartz and fastened vertically atop the vacuum manifold. This eliminated the perpendicular joint that connected the earlier tube with the manifold. It was our calculation that this new design would tolerate higher temperatures with less risk of breakage.

THE EXPERIMENT

The 2011 test employed the same inputs as the 2009 experiments, with several new features. The first modification was the creation of a small recess at the center of the anode. The recess facilitated a more secure placement of test material in the tube. The recess helped confine the test material to the reaction zone. As was the case in 2009, a lead insert was placed in the center of the copper anode. The lead insert consisted of a lead slug approximately 0.25-inch diameter pressed into a 0.25-inch by 0.25-inch drilled hole. One piece of lithium was centrally placed atop the lead insert. The lithium was surrounded by sulfur pieces. Electrode separation was adjusted to a minimum value with just enough clearance to reduce shorting. Typically this was in the range of 0.30 to 0.60 inches.

The second modification was the use of neon as a fill gas to strike plasma before admitting oxygen as the catalyst. (Oxygen was the sole fill gas in the 2009 tests.)

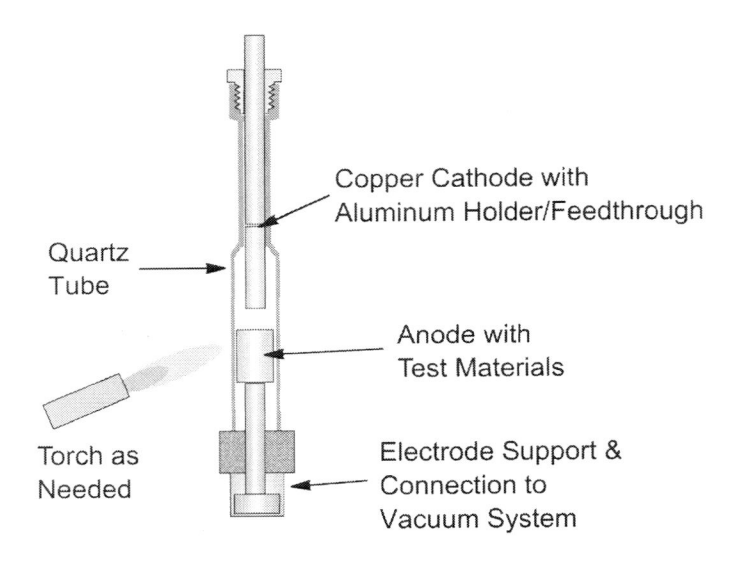

General QR tube configuration showing copper cathode assembly, copper anode, quartz tube, and vacuum system connection. The torch was not used in the test.

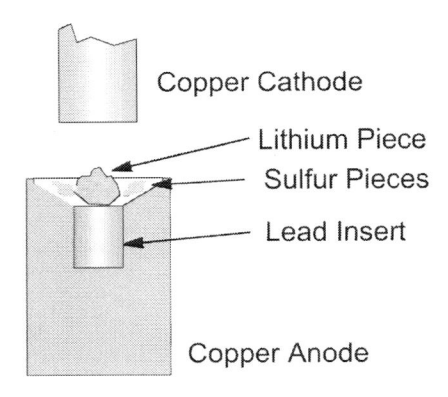

Test 1, Sept 27, 2011

Electrode configuration and test material.

In the worksheet that I drafted prior to the test, I summarized the protocol as follows:

Inputs:
Copper electrodes
Lead insert in anode
Lithium test material
Sulfur test material
Neon fill
Oxygen fill

Procedure:
1. Insert is placed into the anode.
2. Measured quantity of test materials are placed in recess.
3. Glass/quartz tube is placed over the anode assembly.
4. Cathode is inserted into the tube and secured at the desired separation from the anode.
5. Fill with neon to 2 torr.
6. Strike plasma using direct current (DC).
7. Heat reaction zone with hand torch (optional: this was not done in the experiment).
8. Maintain conditions for 2 minutes.
9. Admit oxygen fill to 6 torr. Continue until reaction noticeably slows or tube is in danger of breaking.
10. Disconnect power and allow sample to cool.

The test proceeded in real time as follows (keep in mind that the data points are approximate.) The tube was pumped down to vacuum. Neon was admitted at the start to 2 torr. At 1 minute in, the torr reading was 3.0, while power supply readings measured 53 volts and 6.95 amps. The inside of the tube was glowing red-purple, with what appeared to be the color of neon plasma.

At about 2 minutes, the readings were as follows: 3.0 torr, 70 volts, and 5.63 amps. Intense heat was generated at this point; so that the test materials appeared to be melting. Oxygen was admitted between minute 3 and 4, and the torr reading went up to approximately 8.28. Following the oxygen fill, the test material began glowing a ruby red, the characteristic color of lithium plasma.

At around 6 minutes, there was concern that the tube had failed. Power was disconnected. Thirty seconds later it was decided that the tube was still viable, and the decision was made to admit fresh oxygen and fire up the tube once again. At this point the tube began glowing blue-green. Between 8 to 9 minutes, the power alternated between 45-55 volts and 6.65 and 7.5 amps.

After 10 minutes, conditions in the tube appeared to stabilize. Voltage hovered around 77 and amps around 5.4. The test finished after a total time of approximately 14 minutes with a 3.5 torr reading. Upon conclusion of the experiment, the electricity was turned off, the vacuum pumps disconnected, and the samples allowed to cool.

(For readers who would like to develop their own play-by-play analysis, go to www.CoolFusion.org to see a real time video of the experiment.)

Woody Johnson observes the vacuum discharge test.

> ## The QR Power Supply
>
> The DC (Direct Current) power supply is based on a microwave oven transformer. These transformers are current limited. With a 110 volt mains input the output will be about 3500 volts. As the transformer is loaded, the output voltage will become reduced to maintain a constant output current.
>
> The "hot" output of the transformer is connected to a diode to provide a half-wave rectified output. Since the full output power of the transformer (about 1000 watts) could quickly overheat the QR experiment tube, the primary is ballasted with a high current series resistance consisting of a 600-watt heater element in parallel with a 150-watt incandescent lamp.
>
> For monitoring the power supply we did not measure the actual tube voltage or current through the tube. Instead we monitored the primary voltage and the current into the primary. Voltage was measured with a digital volt meter (DVM) and the current with a commonly available "Kill-A-Watt" meter.
>
> With no load on the transformer, the primary voltage would be close to the mains voltage (118 volts rms is normal here) with the current in the 3-4-amp range. With the tube in an arc condition (high current, low discharge voltage) the primary voltage would drop and the current would increase. The typical arc voltage would be in the 1-volt range vs. glow discharge in the range of a few hundred volts. Note that when the tube is conducting, the voltage across the tube will always be significantly lower than the approx. 3500-volt open circuit voltage. Multiplying the voltage times the current, about 300 V-A in most cases, may approximate power being delivered to the tube. —Steve Hansen, QR Vacuum Consultant

RESULTS & ANALYSIS

Two sets of samples were retrieved for testing: the lead-tipped copper anode with lithium-sulfur residue in its center, the copper cathode, and the quartz tube itself, which contained residue on its inner surface. The anode, cathode, and inside of the tube had undergone noticeable changes during the experiment. The samples were carefully packaged and sent to New Hampshire Materials Laboratory for ICP analysis. The Test Report came back on October 14, 2011 (NHML File Number 28929) and is reproduced below.

>
> New Hampshire
> MATERIALS
> Laboratory, Inc.
>
> TEST REPORT
>
> October 14, 2011
> File Number: 28929
> Mr. Edward Esko
> Quantum Rabbit LLC
>
> Overview:
> Samples Received: QR test samples
> Work Requested: Elemental analysis
> Sample Disposition: Consumed by analysis
>
> Analysis Results:
>
> | Test 1 | Anode/Residue | Cathode/Tube |
> |---|---|---|
> | Potassium | 181 ppm | <0.01 ppm |
> | Gold | 252 ppm | <0.01 ppm |
> | Sample weight (gm) | 0.2083 | 1.7429 |
>
> Prepared by:
> Timothy M. Kenney
> Director of Laboratory Services

The 2011 results paralleled the results of 2009. Two of the three anomalous metals that appeared in 2009 were detected in 2011: potassium (K) at 181 ppm and gold (Au) at 252 ppm. Germanium, the third anomalous metal to appear in 2009, was not detected in the 2011 test. A comparison between the 2009 test and the 2011 experiment is presented below, followed by an analysis of each of the 2011 test components taken from their Certificates of Analysis.

Comparison of 2011 and 2009 Experiments		
Element	Starting Concentration (ppm)*	Final Concentration (ppm)**
2011		
Au	<0.51	252
K	<1.615	181
*2009****		
Au	<0.5	174
K	36.610	750

*Certificate of Analysis provided by Alfa Aesar; additional 2009 value for K provided by M & M Glassblowing.
**ICP Analysis by New Hampshire Materials Laboratory.
***Test 1 conducted on July 30, 2009. Au and K appeared in two subsequent tests on July 30.

Analysis of 2011 Inputs*		
Input	K (ppm)	Au (ppm)
Copper rod (anode) Stock: 12754/Lot: J22P23	<0.01	<0.01
Copper rod (cathode) Stock: 10156/Lot: E18U005	<0.005	<0.5
Lead pellet Stock: 43415/Lot: A10X038	ND**	ND
Lithium piece Stock: 10773/Lot: H17W051	<1	ND
Sulfur pieces Stock: 10755/Lot: A06X003	ND	ND
Quartz tube***	0.6	ND

*Certificate of Analysis provided by Alfa Aesar.
**None detected
***Analysis provided by M & M Glassblowing.

The 2011 results raise the question that was raised in 2009, namely, where did the anomalous metals come from? In the first "Anomalous Metals" article, I presented three possible sources for the anomalies: 1) contamination, or the presence of anomalies in test materials prior to experiment; 2) concentration, or the gathering of anomalous elements in the tested region; and 3) transmutation, or the formation of anomalies through low energy nuclear reactions. The third possibility, that the anomalies appeared through a process of low energy transmutation, is not accepted by modern science.

To glean more background, we asked NHML to do ICP analysis on a control sample of copper taken from the actual batch used in the test, together with a sample of lead from the test batch. The samples were sent to NHML at the end of October and the Test Report, File 29008 presented below, came back on November 3.

New Hampshire
MATERIALS
Laboratory, Inc.

TEST REPORT

November 3, 2011
File Number: 29008

Overview:
Samples Received: (1) Lead Pellet, (2) Copper Rod
Work Requested: Elemental analysis for Au
Sample Disposition: Consumed by analysis

Analysis Results:

Sample	Au Content
Lead Pellet	<1 ppm
Copper Rod	<1 ppm

Prepared by:
Timothy M. Kenney
Director of Laboratory Services

I also requested that each of our vendors reply to the question of whether their equipment or handling of our samples could have led to "contamination" with Au. Steve Hansen, our vacuum system designer and consultant replied: "I am not aware of any way that Au could be introduced. Au is not part of any portion of the system and nothing in the system has been exposed to Au, save the Au that has appeared in the tests. None of the machine tools have Au as a component, etc."

Wayne Martin, of M & M Glassblowing, who fabricated the vacuum tube used in the test answered: "Regarding the Au, I don't see any trace amounts listed in the chemical composition of any of the glasses used to make the tubes. So I guess I can't explain it."

Tim Kenney, who conducted the ICP analysis of our samples at New Hampshire Materials Lab stated: "We rarely deal with Au, and had nothing else in house when your samples were analyzed. Accordingly, the detected Au should not have come from us."

SUMMARY

In the first "Anomalous Metals" article I offered a possible pathway for the transmutation process: "Placing the lithium-sulfur test material in contact with the lead anode, pumping down to vacuum, admitting oxygen as a catalyst and electrifying the tube, may have caused the electrical repulsive force existing between two positively charged nuclei—lithium and sulfur—to become neutralized.... thus allowing the centripetal Casimir force to cause lithium-7 nuclei to fuse with sulfur-32 to produce potassium-39. As fusion took place, a fission reaction occurred in which the lead anode fractured and surrendered nuclei of lithium ($_{204}Pb - _7Li \rightarrow {_{197}}Au$.)"

In other words, according to the quantum conversion hypothesis, the process of cool fusion, in which lithium and sulfur fuse to form potassium, initiates a process of cool fission, in which nuclei of lithium are subtracted from nuclei of lead: Li + S → K (*fusion reaction*) followed by Pb → Li + Au (*fission reaction*).

Such quantum events, if they are eventually proven to occur at all, seem to be taking place in the realm of subatomic particles, thus the purported transmutation products are recorded in parts per million, not in ounces or grams.

Spent electrode with test material.

    Because these events have so far been limited to a minute scale, the specter of contamination, always lurking in the background, is difficult to rule out, our best efforts notwithstanding. Whether or not such quantum events can be replicated on a scale large enough to discount contamination remains to be seen. It also remains to be seen whether such events can be made to occur with regularity and precision, or like the events observed here, occur with a certain degree of randomness and unpredictably.

    In the realm of theory, one scientific reviewer—a trained physicist—has suggested I not attempt to explain these (and other) possible reactions in terms of the Casimir force. As he stated in an E-mail: "It would be better to say that currently there is no explanation for the suggested transmutation reactions, but if transmutation could be confirmed it will be necessary to revise the existing laws of physics."

    However, the phenomenon known as "quantum tunneling" may offer an alternative to the Casimir explanation. In my letter "Quantum Tunneling and the Quantum Rabbit Effect," published in *Infinite Energy* (July 2010) and reprinted in *Cool Fusion*, I propose that under certain conditions, such as those present in QR vacuum tubes, nuclei act more like waves and less like particles. In this model, the nucleus behaves not as a classical particle but as a quantum wave that obeys the laws of quantum tunneling followed by electron waves.

If quantum tunneling applies to nuclei in the way it applies to electron wave-packets, it could explain how a small fraction of positively charged nuclei are able to tunnel through, go around, or somehow breach the Coulomb barrier and achieve fusion with other positively charged nuclei, thus producing the anomalies noted by QR.

The reviewer raises another question: "You have to put a huge amount of energy into the reaction in order to split Pb-204 into Li-7 and Au-197. Where would that energy come from?" Perhaps the energy for the fission reaction comes from the fusion reaction in which lithium combines with sulfur to form potassium. In other words, in this experiment, the fusion of Li-7 and S-32 may have produced a quantum of energy sufficient to trigger the fission of lead into the two lighter elements suggested above.

Source: Edward Esko, "Anomalous Metals Part II," *Infinite Energy* Issue 103, May/June 2012.

14. In Search of the Platinum Group Part II

ABSTRACT

In a study funded by the New Energy Foundation and conducted at Quantum Rabbit (QR) lab in Owls Head, Maine, on September 27, 2011, independent analysis of test samples by inductively coupled plasma spectroscopy (ICP) revealed the anomalous appearance of aluminum (Al)), scandium (Sc), and selenium (Se). The vacuum discharge test employed a copper cathode, pure boron and sulfur test material, and a pure copper anode at the center of which an insert of pure bismuth had been pressed. Scientific grade neon was added to the vacuum tube to strike plasma, followed by a catalyst of pure oxygen. Although it is possible that test materials were contaminated, the appearance of these anomalies raises the possibility of low energy transmutation.

BACKGROUND

The 2011 test followed a series of tests conducted at the QR lab in Owls Head on July 30, 2009, and described in my paper "Anomalous Metals in Electrified Vacuum" published in *Infinite Energy*, Issue 99.

In the 2009 tests, a lead insert was pressed into a copper anode. A copper cathode was inserted into one end of the vacuum tube, followed by pure lithium and sulfur test material. The lead-tipped anode was then inserted, the tube pumped down to vacuum, and pure oxygen admitted to approximately 3.5 torr. An electric arc was struck and a glow discharge with the characteristic color of lithium was produced.

ICP analysis of test samples revealed the anomalous presence of germanium (Ge) at up to 3196 ppm (parts per million); potassium (K) at up to 750 ppm; and gold (Au) at up to 174 ppm. Once again, aside from possible minute traces (mostly below the limits of detection) listed in the Certificates of Analysis of test components, none of the anomalous elements was introduced into the experiment.

In the article I discuss the possibility that the anomalies were produced through a process of low energy transmutation, specifically, a cool fission reaction (Pb − Li → Au) triggered by a cool fusion reaction (Li + S → K).

The 2011 experiment was designed to test the cool fusion → cool fission hypothesis, this time using bismuth rather than lead, and boron in place of lithium, the idea being to see if it would be possible to fission bismuth-209 into platinum-198:

$$_{209}Bi \rightarrow {}_{11}B + {}_{198}Pt$$

This was QR's third attempt at achieving platinum group metals. The first, conducted in September 2008, saw the anomalous appearance of palladium (Pd) on a zinc (Zn) anode, a possible confirmation of the formula: $_{68}Zn + {}_{34}S \rightarrow {}_{102}Pd$. The experiment ("Appearance of Palladium on a Zinc Anode) was published in *Infinite Energy* Issue 87, September/October, 2009. The second effort, a series of experiments designed to produce ruthenium (Ru), was tried in December 2009 and was inconclusive (see "In Search of the Platinum Group Metals, " *Infinite Energy*, Issue 92, July/August 2010.)

VACUUM TUBE DESIGN

In the 2008 and 2009 tests, a vacuum tube made of borosilicate was positioned horizontally atop the vacuum manifold and attached to the manifold by a pump out port. The tube contained a quartz midsection where the reactions took place. The 2011 tube was made entirely of quartz and fastened vertically atop the vacuum manifold. This eliminated the perpendicular joint that connected the earlier tube with the manifold. It was our calculation that this new design would tolerate higher temperatures with less risk of breakage (see illustration of tube design in Chapter 13.)

THE EXPERIMENT

In addition to the vertical tube, the 2011 test differed from the earlier experiments in several respects. The first was the design of a small recess at the center of the anode. The recess facilitated a more secure placement of test material in the tube.

The recess helped confine the test material to the reaction zone. To create the insert, small round beads of pure bismuth, 1-5 mm in size, were poured into a 0.25-inch by 0.25-inch diameter hole and pressed to bind them as one. Boron powder, in the form of tiny crystals -4+40 mesh, was sprinkled atop the bismuth at the center of the anode. Sulfur pieces were then added. Electrode separation was adjusted to a minimum value with just enough clearance to reduce shorting. Usually this was in the range of 0.30 to 0.60 inches.

The second modification was the use of neon as a fill gas to strike plasma before admitting oxygen as the catalyst. (Oxygen was the sole fill gas in the earlier tests.) In the worksheet drafted prior to the test, I summarized the procedure as follows:

Inputs:
Copper electrodes
Bismuth insert in anode
Boron test material
Sulfur test material
Neon fill
Oxygen fill

Procedure:
1. Insert is placed into the anode.
2. Measured quantity of test materials are placed in recess.
3. Glass/quartz tube is placed over the anode assembly.
4. Cathode is inserted into the tube and secured at the desired separation from the anode.
5. Fill with neon to 2 torr.
6. Strike plasma using direct current (DC).
7. Heat reaction zone with hand torch (optional: this was not done in the experiment).
8. Maintain conditions for 2 minutes.
9. Admit oxygen fill to 6 torr. Continue until reaction noticeably slows or tube is in danger of breaking.
10. Disconnect power and allow sample to cool.

In the lab, the test lasted for approximately 10 minutes. Power remained fairly constant at around 50 volts and 7 amps. From the initial 2 torr pressure, the alternating neon and oxygen fill averaged 3.5 torr. At one point, the pressure was increased to 11.9 torr to generate a stronger arc, and then returned to 3.5 torr. Upon completion of the experiment, the power was disconnected and the test sample allowed to cool.

RESULTS

Two sets of samples were retrieved for testing: 1) the copper anode with the bismuth insert and boron-sulfur residue at the center, the copper cathode; and 2) and the quartz tube itself, which contained residue on its inner surface. The anode, cathode, and inside of the tube had undergone noticeable changes during the experiment. The samples were carefully packaged and sent to New Hampshire Materials Laboratory for ICP analysis. The Test Report came back on October 14, 2011 (NHML File Number 28929) and is reproduced on the following page.

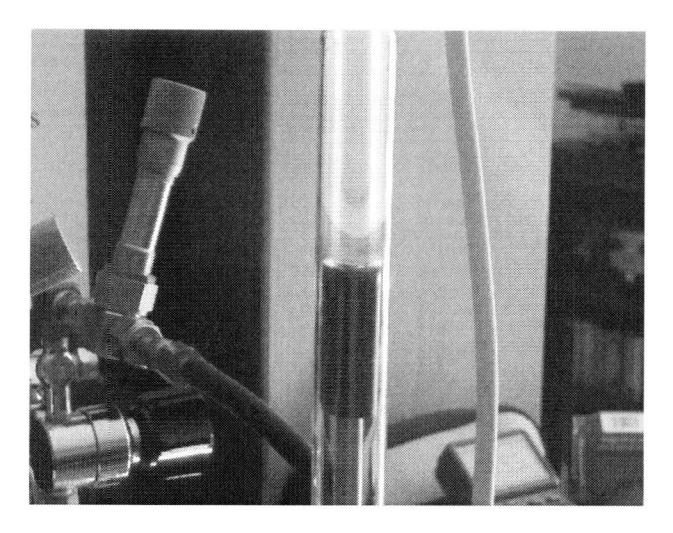

The QR tube with boron-sulfur test material (center) atop the bismuth-filled anode (below.)

ICP analysis revealed the anomalous appearance of three elements: scandium (Sc) at 18 ppm, aluminum (Al) at 342 ppm, and selenium at 12 ppm. Platinum was either not present, or was present in quantities too small to be detected.

A comparison of the quantities of scandium, aluminum, and selenium reported in the Certificates of Analysis of the test materials (Starting Concentration), followed by the quantities of these elements discovered by ICP following the experiment (Final Concentration), appears after the Test Report.

Also presented is a table presenting a breakdown of the quantities of the anomalous elements found in each experimental input as reported in the Certificates of Analysis provided by Alfa Aesar, supplier of the materials used in the test. The analysis of the quartz tube material, provided by M & M Glassblowing, the tube fabricator, is also shown.

New Hampshire
MATERIALS
Laboratory, Inc.

TEST REPORT

October 14, 2011
File Number: 28929
Mr. Edward Esko
Quantum Rabbit LLC

Overview:
Samples Received: QR test samples
Work Requested: Elemental analysis
Sample Disposition: Consumed by analysis

Analysis Results:

Test 3	Anode/Residue	Cathode/Tube
Platinum	<0.01 ppm	<0.01 ppm
Scandium	3 ppm	15 ppm
Aluminum	62 ppm	280 ppm
Selenium	<0.01 ppm	12 ppm
Sample weight (gm)	0.3342	0.3334

Prepared by:
Timothy M. Kenny
Director of Laboratory Services

ANALYSIS

In previous papers I discussed three possible explanations for anomalies such as these: 1) contamination, or the presence of anomalies in test materials prior to experiment; 2) concentration, or the congregation of anomalous elements in the tested region; and 3) transmutation, or the formation of anomalies through low energy nuclear reactions. In this paper, we shall limit ourselves to the third possibility.

Anomalies in the Sept. 27 Test		
Element	Starting Concentration (ppm)*	FinalConcentration (ppm)**
Sc	<0.1012	18
Al	16.143	342
Se	< 0.3087	12

*Certificate of Analysis provided by Alfa Aesar; quantities below detection limits included in totals.
**ICP Analysis by New Hampshire Materials Laboratory.

Analysis of Inputs*			
	Sc ppm	Al ppm	Se ppm
Copper rod (anode) Stock: 12754/Lot: J22P23	<0.001	0.004	<0.01
Copper rod (cathode) Stock: 10156/Lot: E18U005	<0.0002	<0.021	<0.005
Bismuth beads Stock: 38619/Lot: B26W020	<0.1	<0.1	<0.1
Boron powder Stock: 10112/Lot: F05U009	ND**	0.018	<0.0037
Sulfur pieces Stock: 10755/Lot: A06X003	ND	ND	<0.1
Quartz tube***	ND	16	ND

*Certificate of Analysis provided by Alfa Aesar.
**None detected
***Analysis provided by M & M Glassblowing.

In our quantum conversion hypothesis, the process of cool fusion can be used to trigger a process of cool fission. In 2009, we tested this hypothesis with lead (atomic number 82) with intriguing results. In the 2011 tests, we were attempting to test the hypothesis with bismuth (atomic number 83) the element that follows lead on the periodic table, and the heaviest of the non-radioactive elements.

The goal in 2011 was to subtract boron-11 from bismuth-209 (*fission reaction*) to produce platinum-198 ($_{209}Bi \rightarrow {}_{11}B + {}_{198}Pt$). In order to accomplish this, a reaction in which boron-11 fused with oxygen-32 and/or sulfur-34 would be required. In theory, the fusion reactions would result in the creation of aluminum-27 ($_{11}B + {}_{16}O \rightarrow {}_{27}Al$) and/or scandium-45 ($_{11}B + {}_{34}S \rightarrow {}_{45}Sc$).

If all three reactions took place as predicted, in addition to aluminum and scandium, the test sample would have contained platinum. As we can see from the Test Report, platinum was not found in detectable quantities, even though the other predicted metals were noticed. We can interpret this to mean that the fusion reactions were either not strong enough to fission bismuth, or if they were strong enough, the resulting platinum was too insignificant to detect. Meanwhile, as predicted, a secondary reaction seems to have taken place. A small portion of the copper-65 in the electrodes may have fused with boron-11 to form selenium-76: $_{65}Cu + {}_{11}B \rightarrow {}_{76}Se$. [1]

The prevalence of the various isotopes in the raw test materials may explain why certain anomalies seem to be more prevalent than others in the finished samples.

In the reaction $_{11}B + {}_{16}O \rightarrow {}_{27}Al$, we see that boron-11 makes up 80.1% of naturally occurring boron isotopes, and oxygen-16 makes up 99.8% of natural oxygen isotopes. In contrast, in the reaction $_{11}B + {}_{34}S \rightarrow {}_{45}Sc$, sulfur-34 comprises only 4.2% of the naturally occurring isotopes. At 99.8%, oxygen-16 is slightly more than 23 times more widespread amongst oxygen atoms than sulfur-34 is amongst sulfur atoms, thus, the fusion of oxygen-16 with boron-11 yielded 342 ppm aluminum-27, a little more than 20 times the 18 ppm scandium-45 produced by the hypothetical fusion of boron-11 and sulfur-34. There was far more oxygen-16 available for the reaction than there was sulfur-34, a possible explanation for the higher incidence of aluminum-27 in the result.

Footnotes:

[1] The test was repeated on September 27 with similar results (Test 4 NHML File Number 2892). Scandium was reported at 9 ppm, aluminum at 164 ppm, and selenium at 5 ppm. In addition, lithium was introduced in that experiment. Accordingly, sodium was reported by ICP analysis at 50 ppm; a result of possible fusion of lithium-7 with oxygen-16 ($_7Li + {}_{16}O \rightarrow {}_{23}Na$.)

Source: Edward Esko, "In Search of the Platinum Group Part II," *Infinite Energy* Issue 104, July/August 2012.

15. LENR-Induced Transmutation of Nuclear Waste

ABSTRACT

Quantum Rabbit (QR) research on the low energy fusion and fission (low energy nuclear reactions, or LENR) of various elements indicates possible pathways for applying that process to reducing nuclear materials. In a New Energy Foundation (NEF)-funded test conducted at Quantum Rabbit lab in Owls Head, ME, QR researchers initiated a possible low energy fission reaction in which $204Pb$ fissioned into $7Li$ and $197Au$ ($204Pb \rightarrow 7Li + 197Au$) [1]. This reaction may have been triggered by a low energy fusion reaction in which $7Li$ fused with $32S$ to form $39K$ ($7Li + 32S \rightarrow 19K$.) These results confirmed earlier findings showing apparent low energy fusion and fission reactions [2]. Moreover, subsequent research with boron indicates apparent low energy fusion reactions in which boron fuses with oxygen to form aluminum and with sulfur to form scandium [3]. At the same time, the QR group has achieved what appear to be low energy transmutations of carbon using carbon-arc under vacuum and in open air [4]. The research group at QR believes these processes can be adapted to accelerate the natural decay cycle of uranium-235, plutonium-239, radium-226, and the fission products cesium-137, iodine-129, technetium-99, and strontium-90 with the long-term potential of reducing the threat posed by radioactive isotopes to human health and the environment.

Figure 21. Owners and Operators of U.S. Civilian Nuclear Power Reactors Inventories by Material Type as of End of Year, 2005-2009

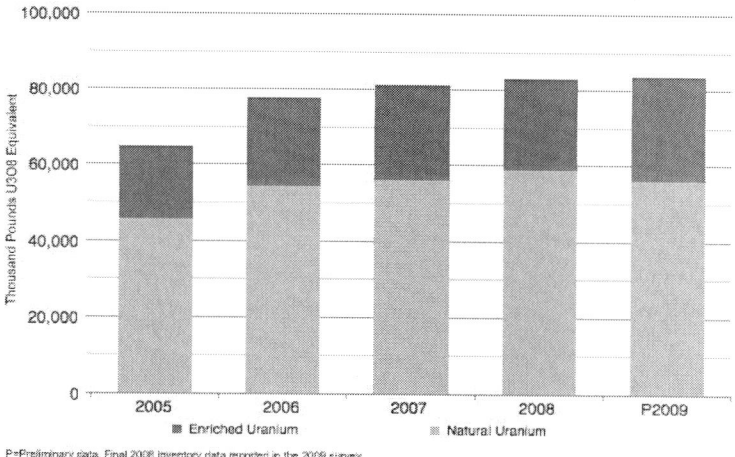

P=Preliminary data. Final 2008 inventory data reported in the 2009 survey.
Source: U.S. Energy Information Administration, Form EIA-858 "Uranium Marketing Annual Survey" (2005-2009).

Figure 1: Inventory of natural and enriched uranium in U.S. nuclear reactors. Enriched uranium at top; natural uranium at bottom.

Uranium-235

In addition to the uranium stored at nuclear reactors (the U.S. inventory is shown in Figure 1,) there are about 2,000 tons of highly enriched uranium in the world, produced mostly for nuclear weapons, naval propulsion, and smaller quantities for research reactors. The half-life of uranium-235 is more than 700 million years. The first step in this process, the alpha decay of uranium-235 into thorium-231 consumes the bulk of this enormous span of time. The half-lives of the isotopes that follow thorium-231 total approximately 33,000 years with the stable isotope lead-207 as the conclusion of the process.

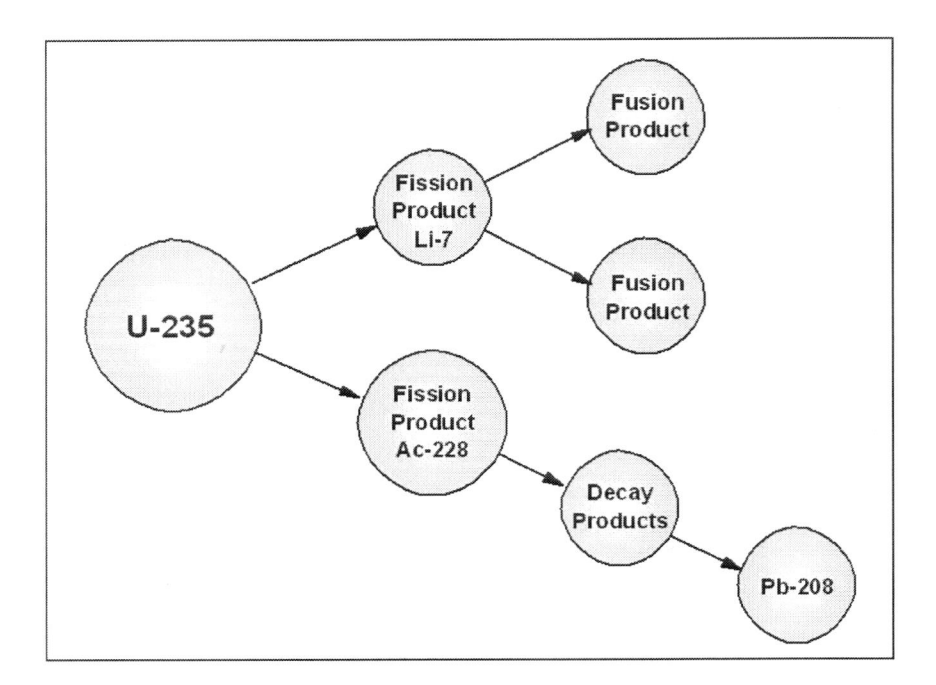

Figure 2: LENR-induced transmutation of uranium-235.

The QR research indicates it may be possible to intervene in the decay cycle of uranium in order to reduce the amount of time needed to achieve its transmutation into lead. The most obvious window for intervention is at the beginning of the cycle, by inducing uranium-235 to fission into one of the lighter isotopes in the radioactive decay chain. We propose using lithium, the catalyst element in the studies cited above, as the catalyst for the following low energy fission reaction:

$_{235}U \rightarrow \, _{7}Li + \, _{228}Ac$
Uranium-235 → lithium-7 + actinium-228

According to this hypothesis, the low energy fusion of lithium with sulfur, resulting in potassium, triggers the low energy fission of uranium into lithium and actinium. The low energy fusion reaction can be written as follows:

$_{7}Li + \, _{32}S \rightarrow \, _{39}K$
Lithium-7 + sulfur-32 → 39K

These reactions are summarized in Figure 2. If achieved, they set in motion the natural decay cycle beginning with actinium-228 and ending with lead-208 shown in Figure 3. Note that the low energy nuclear reaction (LENR) that causes the uranium-235 to fission into actinium-228 results in U-235 being cycled downstream into the natural decay chain of thorium-232 [5].

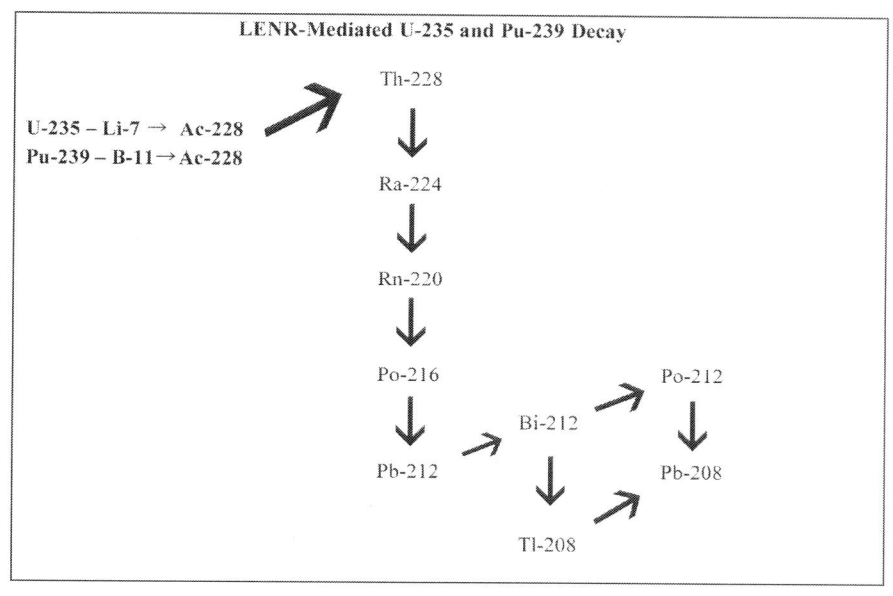

LENR-Mediated U-235 and Pu-239 Decay

U-235 – Li-7 → Ac-228
Pu-239 – B-11 → Ac-228

Th-228 ↓ Ra-224 ↓ Rn-220 ↓ Po-216 ↓ Pb-212 ↗ Bi-212 ↓ ↗ Tl-208 ↗ Pb-208 ↑ Po-212

Figure 3. Accelerated Decay Series: U-235 and Pu-239. Downward arrows represent alpha decay; upward arrows beta decay.

If actinium-228 is produced as predicted, and the natural decay-cycle indicated in Figure 2 set in motion, the half-life of uranium-235 is compressed from over 700-million years to slightly over 1.9 years. The process is summarized in the formula:

$$^{235}U \rightarrow {}_7Li + {}_{228}Ac \rightarrow {}_{208}Pb$$

Uranium-235 → lithium-7 + actinium-228 (thorium-232 decay cycle) → lead-208

Plutonium-239

There is a significant amount of Plutonium-239 in spent nuclear fuel (Table 1).

Composition of 1 Metric Ton of Spent Nuclear Fuel	
	Fission Products
955.4 kg U	10.1 kg lanthanides
8.5 kg Pu (5.1 kg $_{239}$Pu)	1.5 kg $_{137}$Cs
0.5 kg $_{237}$Np	0.7 kg $_{90}$Sr
1.6 kg Am	0.2 kg $_{129}$I
0.02 kg Cm	0.8 kg $_{99}$Te
38.4 kg fission products	0.006 kg $_{79}$Se
	0.3 kg $_{135}$Cs
	3.4 kg Mo isotopes
	2.2 kg Ru isotopes
	0.4 kg Rh isotopes
	1.4 kg Pd isotopes

Table 1. Source: James Laidler, Development of Separations Technologies Under the Advanced Fuel Cycle Initiative. Report to the ANTT Subcommittee, December 2002.

It may also be possible to use LENR to compress the decay cycle of plutonium-239. Plutonium-239 has a half-life of 24,000 years. As mentioned in the Abstract, the QR research group achieved promising results with the low energy fusion of boron. A series of experiments for the possible reduction of plutonium-239 similar to the QR experiments can be designed using boron as the catalyst element. The low energy fission reaction we propose testing is as follows:

$_{239}$Pu → $_{11}$B + $_{228}$Ac
Plutonium-239 → boron-11 + actinium-228

This low energy fission reaction is theoretically triggered by several low energy fusion reactions [6]:

$_{11}$B + $_{16}$O → $_{27}$Al

Boron-11 + oxygen-16 → aluminum-27

$_{11}B + {}_{34}S \rightarrow {}_{45}Sc$
Boron-11 + sulfur-34 → scandium-45

Once again, if these LENR are successful in producing actinium-228, like U-235 in the formula described earlier, Pu-239 would be cycled downstream into the thorium-232 decay chain with the end product being the stable isotope lead-208 (Figure 3.) This process can be summarized as follows:

$_{239}Pu \rightarrow {}_{11}B + {}_{228}Ac \rightarrow {}_{208}Pb$
Plutonium-239 → boron-11 + actinium-228 (thorium-232 decay cycle) → lead-208

Radium-226

Contamination by radium-226 continues to be a problem at military installations and other sites around the world. Radium-226 is part of the U-238 decay chain with a half-life of 1,600 years. With LENR, it may be possible to compress this time frame considerably by achieving the low energy fission of Ra-226.

Earlier QR research on carbon-arc may offer a method for achieving this possibility. Numerous LENR have been reported, both in open air and under vacuum [7]. These low energy fusion reactions could possibly be used to prompt the low energy fission of radium-226, compressing the half-life of radium and accelerating the natural decay cycle from more than 1,600 years to approximately 22 years. (Figure 4.)

The low energy fission reaction we propose testing is as follows:

$_{226}Ra \rightarrow {}_{12}C + {}_{214}Pb$
Radium-226 → carbon-12 + lead-214

This low energy fission reaction could possibly be triggered by low energy fusion reactions such as those between carbon and oxygen noted in QR carbon-arc research:

$_{12}C + {}_{12}C \rightarrow {}_{24}Mg$
Carbon-12 + carbon-12 → magnesium-24

$_{12}C + _{16}O \rightarrow _{28}Si$
Carbon-12 + oxygen-16 → silicon-28

$_{12}C + 2(_{16}O) \rightarrow _{44}Ti$
Carbon-12 + 2(oxgyen-16) → titanium-44

$_{12}C + _{32}S \rightarrow _{44}Ti$
Carbon-12 + sulfur-32 → titanium-44

$2(_{12}C + _{16}O) \rightarrow _{56}Fe$ (+ 2 protons)
2(Carbon-12 + oxygen-16) → iron-56 + 2 protons

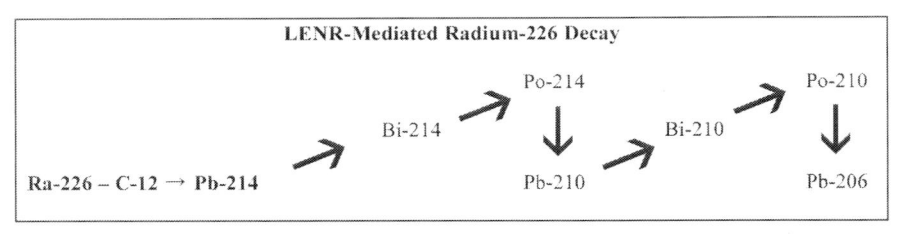

LENR-Mediated Radium-226 Decay

Ra-226 – C-12 → **Pb-214** ↗ Bi-214 ↗ Po-214 ↓ Pb-210 ↗ Bi-210 ↗ Po-210 ↓ Pb-206

Figure 4. Accelerated Decay Series: Ra-226. Downward arrows represent alpha decay; upward arrows beta decay.

Cesium-137

Cesium-137, a product of nuclear fission is a major radionuclide in spent nuclear fuel. It is of major concern for Department of Energy environmental management sites and has a half-life of 30 years. It decays by emitting a beta particle. It's decay product, barium-137m (the "m" is for metastable) stabilizes by emitting an energetic gamma ray with a half-life of approximately 2.6 minutes. It is this decay product that qualifies cesium-137 as a radiation hazard.

The environmental dangers posed by cesium-137 were highlighted by the crisis at Fukushima Daichi reactor in Japan. Writing in the Proceedings of the National Academy of Sciences [8], an international team of scientists described the threat posed by cesium-137:

The largest concern on the cesium-137 (137Cs) deposition and its soil contamination due to the emission from the Fukushima Daiichi Nuclear Power Plant (NPP) showed up after a massive quake on March 11, 2011. Cesium-137 (137Cs) with a half-life of 30.1 y causes the largest concerns because of its deleterious effect on agriculture and stock farming, and, thus, human life for decades. Removal of 137Cs contaminated soils or land use limitations in areas where removal is not possible is, therefore, an urgent issue.

Contamination by cesium-137 was a major problem following the Chernobyl disaster. As John Emsley states [9]:

> Uranium fuel rods in nuclear power stations produce cesium-137. The half-life of cesium-137 is 30 years, which means that it takes over 200 years to reduce it to 1% of its former level. For this reason, an accident at a nuclear power plant can contaminate the environment around for generations, which is why the Chernobyl accident in the Ukraine in 1986 was such an environmental disaster. It released a large amount of radioactive cesium-137 which drifted all over Western Europe, affecting sheep farms as far west as Scotland, Ireland, and Wales, over 1500 miles from the accident. There it was washed to earth by heavy rain and taken up by the roots of plants, thus becoming part of the vegetation that sheep ate.

Using LENR, it may be possible to convert cesium-137 to tellurium-130, a stable non-radioactive isotope, thus redirecting and compressing the cesium-137 decay cycle (Figure 5). The LENR-induced fission formula is as follows:

$_{137}Cs \rightarrow {}_7Li + {}_{130}Te$
Cesium-137 → lithium-7 + tellurium-130

In theory, the low energy fission reaction would be triggered by the low energy fusion of lithium and sulfur:

$_7Li + {}_{32}S \rightarrow {}_{39}K$
Lithium-7 + sulfur-32 → potassium-39

In a separate experiment, cesium-137 may also transmute into neodymium-148 through a low energy fusion reaction:

$_{137}Cs + _{11}B \rightarrow _{148}Nd$
Cesium-137 + boron-11 → neodymium-148

If the fusion reaction can be proven and scaled to production levels, it would then be possible to convert dangerous radioactive waste into a valuable rare earth metal widely utilized today in the magnets in hybrid vehicles.

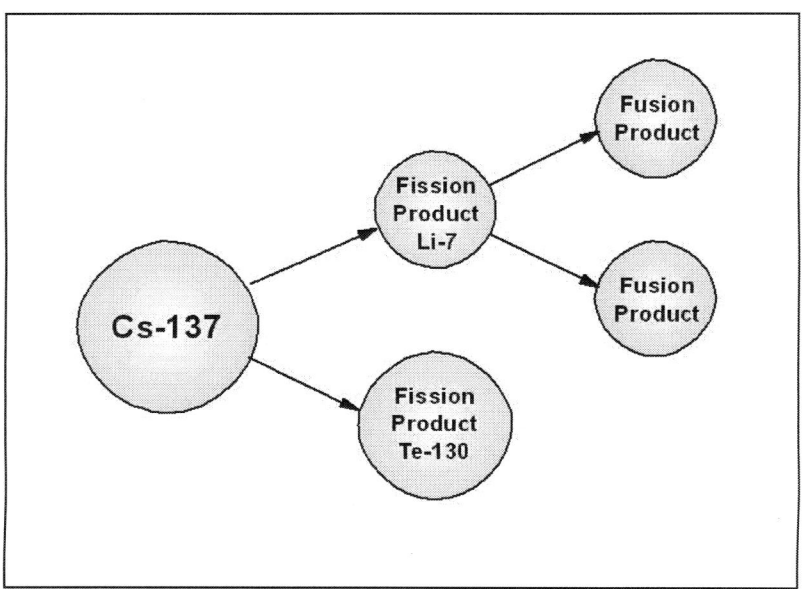

Figure 5: LENR-induced transmutation of cesium-137.

Iodine-129
Iodine-129 is a long-lived isotope of iodine created primarily from the fission of uranium and plutonium in nuclear reactors. It decays with a half-life of 15.7 million years. Significant amounts were released into the atmosphere following nuclear weapons tests in the 1950s and 1960s. Iodine-129 is long-lived and mobile in the environment and is thus of special importance in disposal and management of spent nuclear fuel. It may be possible to compress the natural decay cycle of this radioisotope through the process of low energy fission. The LENR-induced fission reaction is as follows:

$_{129}I \rightarrow {_7}Li + {_{122}}Sn$
Iodine-129 → lithium-7 + tin-122

Once again, according to theory, low energy fusion of lithium and sulfur would serve as the catalyst for the reaction:

$_7Li + {_{32}}S \rightarrow {_{39}}K$
Lithium-7 + sulfur-32 → potassium-39

During the experiment, iodine-129 may also transmute into barium-146 through a simultaneous fusion reaction:

$_{129}I + {_7}Li \rightarrow {_{136}}Ba$
Iodine-129 + lithium-7 → barium-136

Technetium-99

Technetium-99 is radioisotope of technetium that decays with a half-life of 211,000 years to stable ruthenium-99. It is the most significant long-lived fission product of uranium-235. Its high fission yield, relatively long half-life, and mobility in the environment make technetium-99 one of the more problematic components of nuclear waste. There have been releases into the environment from atmospheric nuclear tests, nuclear reactors, and in the late 1990s from the Sellafield plant, which released nearly 1,000 kg into the Irish Sea.

It may be possible to accelerate the half –life of Tc-99 by inducing the following low energy fission reaction:

$_{99}Tc \rightarrow {_7}Li + {_{92}}Zr$
Technetium-99 → lithium-7 + zirconium-92

Once again, in theory, the reaction would be triggered by the low energy fusion of lithium and sulfur:

$_7Li + {_{32}}S \rightarrow {_{39}}K$
Lithium-7 + sulfur-32 → potassium-39

During the experiment, Tc-99 may also transmute into Pd-106 through the following fusion reaction:

$_{99}Tc + {}_7Li \rightarrow {}_{106}Pd$
Technetium-99 + lithium-7 → palladium-106

Strontium-90

Together with cesium-137, strontium-90 is component of spent nuclear fuel. These radioisotopes have an intermediate half-life of about 30 years, the worst range for half-lives of radioactive waste products. Not only are they highly radioactive, but also they have long enough half-lives to last for hundreds of years. Strontium-90 acts like calcium and is taken up by plants and animals and deposited in bones. John Emsley describes strontium-90 as follows [10]:

> Strontium-90 caused a major worldwide pollution concern in the mid-twentieth century, being produced by aboveground nuclear explosions which contaminated the whole planet with it. These tests took place between 1945 and 1963. Strontium-90 is a serious threat because it is one of the most powerful emitters of ionizing radiation and therefore capable of causing serious damage to dividing cells. Its presence was detected in the milk teeth of infants in the 1950s, showing how prevalent it had become, having been washed out the atmosphere on to grassland, to be eaten by cows, and so end up in milk and other dairy products.

It may be possible to remediate strontium-90 through a simple low energy fusion process. The formula is as follows:

$_{90}Sr + {}_{12}C \rightarrow {}_{102}Ru$
Strontium-90 + carbon-12 → ruthenium-102

GUIDELINES FOR METHODOLOGY

The experiments on low energy transmutation cited in the Abstract can serve as a starting point for designing experiments to test the nuclear reduction hypothesis presented in this paper [11].

A vacuum tube similar to that used in the QR low energy transmutation tests and shown in Figure 6 can be considered for the nuclear reduction tests. Because silver is a strong conductor of electricity and a neutron absorber, we propose using it as the anode and cathode material, with other test materials adjusted for each experiment as indicated below. Moreover, silver may react independently with lithium to form tin ($_{109}Ag + _7Li \rightarrow {_{116}Sn}$). This reaction was noted in a previous QR test [12]. Keep in mind that these suggestions are guidelines only, based on previous low energy fusion and fission experiments. They will need to be adjusted in real time based upon further study and experience.

Uranium-235:

$_{235}U \rightarrow {_7Li} + {_{228}Ac} \rightarrow {_{208}Pb}$

Electrodes made of Ag
Test Materials:

1. Uranium insert (thin wafer or foil) in anode
2. Lithium test material
3. Sulfur test material
4. Pure neon/oxygen backfill

Plutonium-239:

$_{239}Pu \rightarrow {_{11}B} + {_{228}Ac} \rightarrow {_{208}Pb}$

Electrodes made of Ag
Test Materials:

1. Plutonium insert (thin wafer or foil) in anode
2. Boron test material
3. Sulfur test material (optional)
4. Pure neon/oxygen backfill

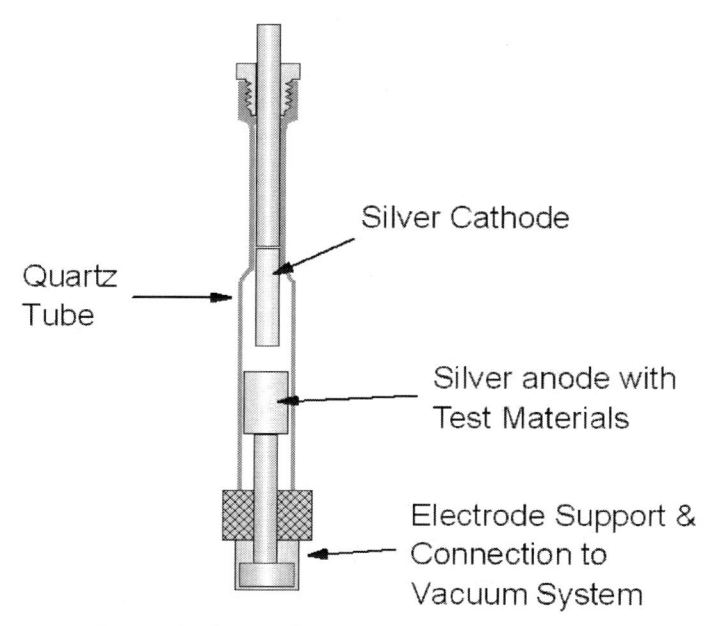

Figure 6. Tube and electrode configuration.

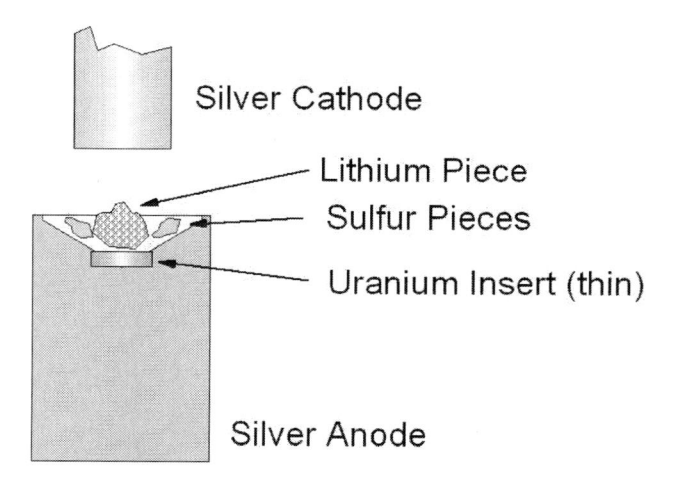

Figure 7. Electrodes and test material suggested for the U-235 experiment.

Radium-226:

$_{226}Ra \rightarrow {}_{12}C + {}_{214}Pb \rightarrow {}_{206}Pb$

Electrodes made of Ag
Test Materials:

1. Radium insert (thin wafer or foil) in anode
2. Carbon (graphite) test material
3. Sulfur test material
4. Pure nitrogen/oxygen backfill*

*Note: Adding nitrogen allows the process to take advantage of potential carbon-nitrogen reactions such as those noted in QR research [13].

Cesium-137:

$_{137}Cs \rightarrow {}_{7}Li + {}_{130}Te$

Electrodes made of Ag
Test Materials:

1. Cesium insert (thin wafer or foil) in anode
2. Lithium test material
3. Sulfur test material
4. Pure neon/oxygen backfill

$_{137}Cs + {}_{11}B \rightarrow {}_{148}Nd$

Electrodes made of Ag
Test Materials:

1. Cesium insert (thin wafer or foil) in anode
2. Boron test material
3. Sulfur test material
4. Pure neon/oxygen backfill

Iodine-129:

$_{129}I \rightarrow {_7}Li + {_{122}}Sn$
$_{129}I + {_7}Li \rightarrow {_{136}}Ba$

Electrodes made of Ag
Test Materials:

1. Iodine inserted in or on anode
2. Lithium test material
3. Sulfur test material
4. Pure neon/oxygen backfill

Technetium-99:

$_{99}Tc \rightarrow {_7}Li + {_{92}}Zr$
$_{99}Tc + {_7}Li \rightarrow {_{106}}Pd$

Electrodes made of Ag
Test Materials:

1. Technetium insert (thin wafer or foil) in anode
2. Lithium test material
3. Sulfur test material
4. Pure neon/oxygen backfill

Strontium-90:

$_{90}Sr + {_{12}}C \rightarrow {_{102}}Ru$

Electrodes made of Ag
Test Materials:

1. Strontium insert (thin wafer or foil) in anode
2. Carbon (graphite) test material
3. Sulfur test material
4. Pure neon/oxygen backfill

Procedure for the Above Experiments:

1. Insert is placed on or into the anode.
2. Measured quantity of test materials are placed in anode recess.
3. Glass/quartz tube is placed over the anode assembly.
4. Cathode is inserted into the tube and secured at the desired separation from the anode.
5. Fill with neon (or nitrogen for Ra-226) to 2 torr.
6. Strike plasma using direct current (D.C.)
7. Admit oxygen fill to 6 torr. Continue until reaction noticeably slows or tube is in danger of breaking (approximately 10-20 minutes.)
8. Disconnect power and allow sample to cool.

CONCLUSION

As of this writing, the problem of nuclear waste disposal remains unsolved. In an op-ed published in the Santa Monica Daily Press [14], Dr. Jeffrey Patterson, former head of Physicians for Social Responsibility (PSR) stated:

> 2011 was a scary year for nuclear reactor sites. The summer floods threatened to encroach on reactors in Nebraska and Iowa, an earthquake and a hurricane happened in quick succession to rattle and flood the East Coast, and the continuing events of the Fukushima-Daichi reactor accident provided harrowing examples of the threats posed to spent fuel at reactor sites. The fate of spent fuel there kept the world on edge for days. It's worth noting that the amount of fuel in vulnerable storage pools in Japan was far less than what is crowded into pools at many U.S. reactors. As we all learned, a loss of coolant could produce a fuel melt and large radiation release.
>
> It wasn't supposed to be this way. Used reactor fuel was to be permanently stored in deep underground repositories, away from floods and other natural hazards. But the solution to the nation's nuclear waste problem has been elusive for decades. Meanwhile, 65,000 metric tons of spent reactor fuel is still looking for a home.

The Blue Ribbon Commission on America's Nuclear Future proposes transferring spent nuclear fuel, now scattered at 70 locations around the U.S. to temporary storage areas, pending selection of more permanent deep geologic repositories. This proposal is not without controversy. As Dr. Patterson states:

> Moving spent fuel around the country is not a risk worth taking. Rather than addressing the problem, an "interim" facility would only relocate it. So what is the best option? Hardened on-site storage of spent fuel. It's safe, cost-effective—and readily available. PSR and over 170 public interest organizations from all 50 states are calling for adoption of this approach. Storing reactor fuel at reactor sites in hardened buildings that can resist severe attacks, such as a direct hit by high-powered explosives or a large aircraft, as is done in Germany, offers the safest and most sensible option until a permanent repository can be found.

These proposals offer opportunities for research on LENR-induced transmutation. Research laboratories could be set up at future on-site hardened facilities or even now at current waste storage sites, as well as at future interim facilities. These laboratories can begin first-round investigation of LENR-induced transmutation. If successful, scale-up can proceed to levels required to reduce the on-site, regional, and global inventory of nuclear waste.

Moreover, LENR-induced transmutation may offer an efficient low-cost alternative to accelerator transmutation of waste (ATW). In 1999, the U.S. Department of Energy's (DOE) Office of Civilian Radioactive Waste Management submitted a report to Congress entitled "A Roadmap for Developing Accelerator Transmutation of Waste (ATW) Technology." Sekazi K. Mtingwa of MIT describes this approach as follows [15]:

> Transmutation means the transformation of one atom into another by changing its nuclear structure. In the present context this means bombarding a highly radioactive atom with neutrons, preferably fast neutrons, from either a fast nuclear reactor or spallation neutrons created by bombarding protons from a high-energy accelerator on a suitable target.

The Oak Ridge National Laboratory (ORNL) is currently investigating methods for accelerator transmutation of nuclear wastes. An article in the ORNL *Review* states [16]:

> Conceived by scientists at Los Alamos National Laboratory, ATW uses a linear accelerator system to produce neutrons for transmutation of excess weapons plutonium and other radioactive DOE wastes, such as technetium-99 and iodine-129. Ultimately, the potential of partitioning and transmutation to waste management is this: If a radioactive waste stream no longer exists, then it poses no radiological hazard. More than anything else, this simple fact has spurred the recent resurgence of interest in partitioning-transmutation technology.

Meanwhile, in the Eurozone, the European nuclear establishment is pressing ahead with a $1.2 billion R & D project to look into high-energy neutron-induced transmutation. The first stage of the project, the setup of a demonstration system known as "Guinevere" that combines a particle accelerator and a nuclear reactor, took place in January 2012 at the Belgian Nuclear Research Center at Mol. A larger version of the reactor system, known as Myrrha (Multipurpose Hybrid Research Reactor for High-tech applications), is scheduled to become operational in 2023. A press release from the World Nuclear Association explains the thinking behind the project [17]:

> Myrrha will be able to produce radioisotopes and doped silicon, but its research functions would be particularly well suited to investigating transmutation. This is when certain radioactive isotopes with long half-lives are made to 'catch' a neutron and thereby change into a different isotope that will decay more quickly to a stable form with no radioactivity. If achievable on an industrial scale, transmutation could greatly simplify the permanent geologic disposal of radioactive waste.

The Quantum Rabbit group estimates that research on LENR-induced transmutation could begin at a fraction of the estimated $1.2 billion startup cost of the Myrrha project. (QR estimates $1.2 million for feasibility study and $12 million to develop a prototype system, amounts that are 0.1% and 1% the cost of Myrrha.)

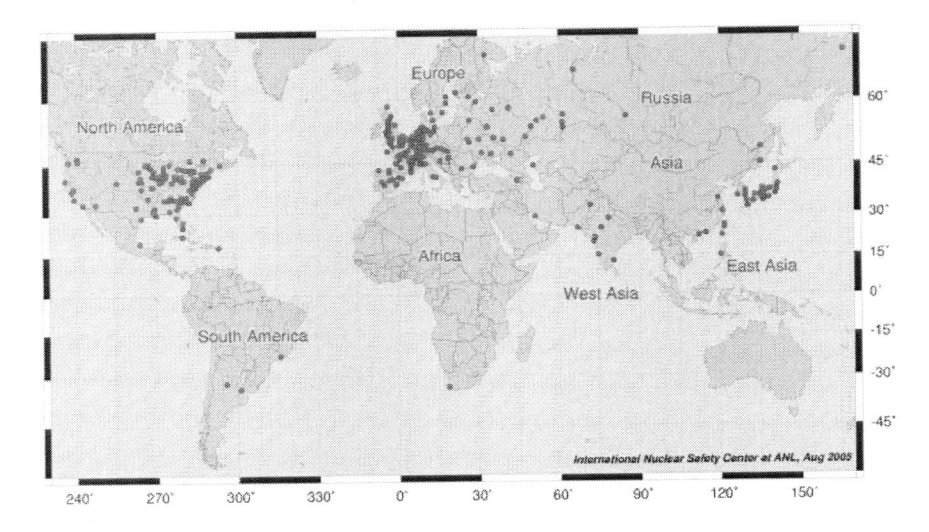

Figure 8: Nuclear power stations around the world.

Rather than a highly centralized billion-dollar processing system, LENR-induced transmutation technology could be distributed to nuclear power stations around the globe (Figure 8) at an affordable cost. The task of nuclear remediation would become the responsibility of the individual power station and thus remain local instead of becoming highly centralized. Also, the amount of power needed to conduct LENR-induced transmutation would be miniscule compared to the power required to operate a particle accelerator and nuclear reactor. At the very least, research on LENR-induced transmutation should proceed on a parallel track to the high-energy neutron-induced transmutation projects currently underway or under consideration in order to determine which approach yields the most promising results.

The author wishes to thank Charles Entenmann for suggesting research on the modification of the radioactive decay chain utilizing low temperature, pressure, and energy.

[1] Esko, Edward, "Anomalous Metals Part II," Infinite Energy, No. 103, 2012.
[2] Esko, Edward and Jack, Alex, Cool Fusion, first edition, Amber Waves, Becket, Mass., USA, pp. 108-118, 2011.
[3] Esko, Edward, "In Search of the Platinum Group Metals Part II," Infinite Energy, No. 104, 2012.

[4] Esko, Edward and Jack, Alex, Cool Fusion, first edition, Amber Waves, Becket, Mass., USA, pp. 61-65, 93-102, 2011.
[5] Argonne National Laboratory, "Human Health Fact Sheet," Figure N.3 Natural Decay Series: Thorium-232, 2005.
[6] Esko, Edward, "In Search of the Platinum Group Metals Part II," Infinite Energy, no. 104, 2012.
[7] Esko, Edward and Jack, Alex, Cool Fusion, first edition, Amber Waves, Becket, Mass., USA, pp. 61-65, 93-102, 2011.
[8] Yasunari, Teppei J., Stohl, Andreas, Hayano, Ryugo S., Burkhart, John F., Eckhardt, Sabine, Yasunari, Tetsuzo, "Cesium-137 deposition and contamination of Japanese soils due to the Fukushima nuclear accident," Proceedings of the National Academy of Sciences, November, 14, 2011.
[9] Emsley, John, Nature's Building Blocks: An A-Z Guide to the Elements, first edition, Oxford University Press, Oxford, England, pp. 82, 2001.
[10] Emsley, John, Nature's Building Blocks: An A-Z Guide to the Elements, first edition, Oxford University Press, Oxford, England, p. 407, 2001.
[11] Refer to Occupational Safety and Health Administration (OSHA) guidelines for the handling of hazardous materials prior to initiating these experiments.
[12] Esko, Edward and Jack, Alex, Cool Fusion, first edition, Amber Waves, Becket, Mass., USA, pp. 84-92, 2011.
[13] Esko, Edward and Jack, Alex, Cool Fusion, first edition, Amber Waves, Becket, Mass., USA, pp. 63, 100, 2011.
[14] Patterson, Jeffrey, "Time to fix our nuclear waste disposal system," Santa Monica Daily Press, January 06, 2012.
[15] Mtingwa, Sekazi K., "Feasibility of Transmutation of Radioactive Isotopes," An International Spent Nuclear Fuel Storage Facility -- Exploring a Russian Site as a Prototype: Proceedings of an International Workshop, The National Academies Press, Washington, D.C., 2005.
[16] Michaels, Gordon E., "Partitioning and Transmutation: Making Wastes Nonradioactive," Oak Ridge National Laboratory Review, Vol. 44, No. 2, 2011.
[17] World Nuclear News, World Nuclear Association, January 11, 2012.

Source: Edward Esko, "LENR-Induced Transmutation of Nuclear Waste," *Infinite Energy* Issue 104, July/August 2012.

APPENDIX

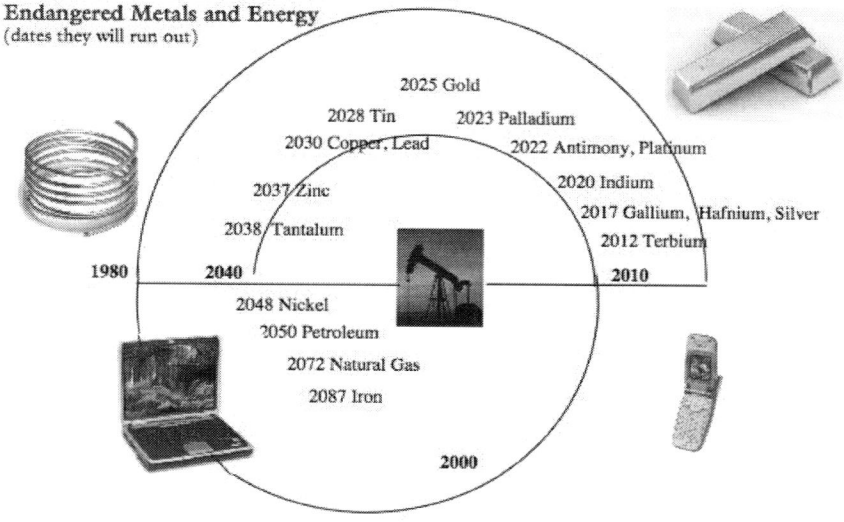

Vanishing Metals

Endangered Metals and Energy
(dates they will run out)

- 2025 Gold
- 2028 Tin
- 2023 Palladium
- 2030 Copper, Lead
- 2022 Antimony, Platinum
- 2037 Zinc
- 2020 Indium
- 2038 Tantalum
- 2017 Gallium, Hafnium, Silver
- 2012 Terbium
- 1980
- 2040
- 2010
- 2048 Nickel
- 2050 Petroleum
- 2072 Natural Gas
- 2087 Iron
- 2000

Vanishing Metals
© 2011 by Alex Jack

Source: *New Scientist, Wall Street Journal,* and the latest market forecasts from world commodity exchanges.

RESOURCES

Quantum Rabbit LLC The Massachusetts Limited Liability Company (LLC) formed in 2005 by Edward Esko, Alex Jack, and Woody and Florence Johnson. For further information, including investment opportunities, contact Edward Esko, 109 Wendell Ave., Pittsfield MA 01201, (413) 442-1360, edwardesko@gmail.com.

CoolFusion.org Web site of Quantum Rabbit with information and articles on Cool Fusion.

Planetary Health Inc. A nonprofit educational organization and sponsor of Amberwaves, a grassroots network established to protect whole grains and other essential foods from genetic engineering, climate change, and other hazards; CERES, The Committee to Explore and Research Energy Solutions; and Amber Waves Press, publisher of *The Element Genome* and other literature. The quarterly *Amberwaves Journal* includes continuing coverage of QR experiments and activities and is $25/year.

For further information, including tax-deductible donations, contact Alex Jack, 305 Brooker Hill Road, Becket MA 01223, (413) 623-0012, shenwa@bcn.net, www.amberwaves.org.

Woodland Energy Co. A green energy company founded by Woodward Johnson and maker of the HUBERT® used in the QR experiments. For further information, contact Woodward Johnson, 200 Bush Hill Rd., Ashburnham MA 01430, (978) 827-5055, woodward1984@gmail.com, www.woodland-energy.com.

New Energy Foundation Sponsor of research in new energy science and technology, including QR experiments. NEF's magazine, *Infinite Energy*, reports on developments in cold fusion, cool fusion, and other promising technologies. Subscriptions are $29.95. For further information, contact: P.O. Box 2816, Concord NH 03302, www.infinite-energy.com.

ABOUT THE AUTHORS

Edward Esko began his study of macrobiotic cosmology, including low energy nuclear transmutations (LENT), in the 1970s with Michio Kushi in Boston. His study included the theory and practice of energy meridians, such as those used in acupuncture, the energetic effects of food, the universal stages of energy transformation, and the universal laws of harmony and balance as they appear in the world of physics and chemistry. In 1978 he traveled to Japan to deepen his understanding of Oriental cosmology and spirituality. He joined the faculty of Kushi Institute in 1980, and began lecturing on practical applications of the cosmology of change as a means to achieve planetary health and peace. In 2004 Edward joined with Alex Jack and Woody Johnson to form Quantum Rabbit LLC, a Massachusetts Limited Liability Company, for the purpose of developing the theory of LENT (low energy nuclear transmutations) presented by Louis Kervran, George Ohsawa, Michio Kushi, and other pioneers. Contact: edwardesko@gmail.com

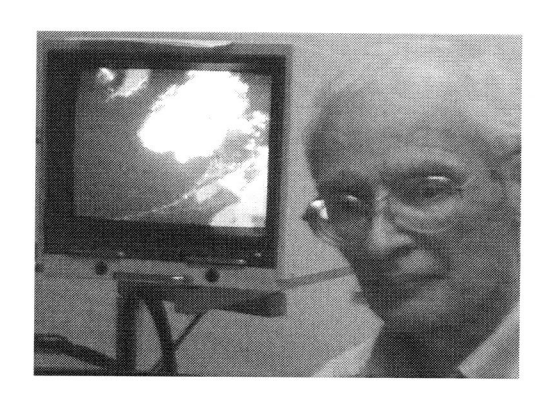

Alex Jack served as a reporter in Vietnam, editor-in-chief of the *East West Journal*, general manager of the Kushi Institute, director of the One Peaceful World Society, and president of Planetary Health. He is the author or editor of many books, including *The Cancer Prevention Diet* with Michio Kushi, *Aveline Kushi's Complete Guide to Macrobiotic Cooking*, *The Mozart Effect* by Don Campbell, and *Buddha Standard Time* by Lama Surya Das. His multi-volume series, *Profiles in Oriental Diagnosis*, explores the creators of the modern mind, including Leonardo, Newton, Descartes, Darwin, and Pasteur. After the Chernobyl nuclear accident, he organized an airlift to the Soviet Union of foods that help protect against radiation. He has spoken at the Zen Temple in Beijing, the Cardiology Center in St. Petersburg, and Shakespeare's Globe Theatre in London. He is a guest lecturer at the Kushi Institute of Europe and the Ohsawa Center in Japan. He lives in the Berkshires. Contact: shenwa@ bcn.net.

Dr. Mahadeva Srinivasan is a nuclear scientist and one of the world's leading authorities on LENR (Low Energy Nuclear Reactions), including transmutation. He was the head of the Neutron Physics Division and an Associate Director of the Physics Group of BARC (Bhabha Atomic Research Center) in Mumbai (Bombay). Since his retirement from BARC in 1997, at the end of a four-decade long research career, Dr. Srinivasan has been living in Chennai, India. He continues to take a keen interest in developments in cold fusion. In February 2011 he chaired the 16th International Conference on Condensed Matter Nuclear Science (ICCF 16) in Chennai. He is the author of "Low Energy Nuclear Reactions: Transmutations" with George Miley and Edmund Storms, in Jay H. Lehr, Steven B. Krivit, and Thomas B. Kingery, editors, *Nuclear Energy Encyclopedia: Science, Technology, and Applications*, John Wiley & Sons, 2011.